XXIIIrd National Conference on Thermodynamics with International Participation

AA007381

IOP Conference Series: Material Science and Engineering
Volume 1290

Galati, Romania
11-13 May 2023

ISBN: 978-1-7138-7969-5
ISSN: 1757-8981

Printed from e-media with permission by:

Curran Associates, Inc.
57 Morehouse Lane
Red Hook, NY 12571

Some format issues inherent in the e-media version may also appear in this print version.

This work is licensed under a Creative Commons Attribution 3.0 International Licence.
Licence details: http://creativecommons.org/licenses/by/3.0/.

No changes have been made to the content of these proceedings. There may be changes to pagination and minor adjustments for aesthetics.

Printed with permission by Curran Associates, Inc. (2023)

For permission requests, please contact the Institute of Physics
at the address below.

Institute of Physics
Dirac House, Temple Back
Bristol BS1 6BE UK

Phone: 44 1 17 929 7481
Fax: 44 1 17 920 0979

techtracking@iop.org

Additional copies of this publication are available from:

Curran Associates, Inc.
57 Morehouse Lane
Red Hook, NY 12571 USA
Phone: 845-758-0400
Fax: 845-758-2633
Email: curran@proceedings.com
Web: www.proceedings.com

TABLE OF CONTENTS

APPLIED THERMODYNAMICS, HEAT AND MASS TRANSFER, FLUID DYNAMICS AND CFD

REFRIGERATION AND HEAT PUMPS, POWER PLANTS AND CHP, INTERNAL COMBUSTION ENGINES, AUTOMOTIVES, ENGINE AND TRANSMISSION, ROAD SAFETY

RENEWABLE ENERGY, ENERGY USE AND STORAGE, ENVIRONMENTAL IMPACT OF ENERGY CONVERSION

IOP Conf. Series: Materials Science and Engineering 1290 (2023) 011001 doi:10.1088/1757-899X/1290/1/011001

Every two years the Department of Thermal Systems and Automotive from the "Dunarea de Jos" University of Galati, Romania in cooperation with the Romanian Society of Thermal Engineering (SRT) and the Romanian Society for Refrigeration and Cryogenics Engineers (AFCR) organizes and hosts the National Conference on Thermodynamics with International Participation. This year, the 23rd edition of the conference, NACOT 2023, took place. The Scientific Advisory Board has selected the topic of the conference and its participants. The conference program included invited talks, onsite participants and online participants from Romania, Poland, Belarus, Greece, Portugal, Tunis, Kazakhstan, Bulgaria, Turkey and Republic of Moldova.

The conference covers a range of topics related to thermodynamics: Applied thermodynamics; Heat and mass transfer; Fluid dynamics and CFD; Refrigeration and heat pumps; Power plants and CHP; Internal combustion engines; Renewable energy (solar, wind, hydro, biomass, etc.); Energy use (industry, building, transportation, desalination, etc.) and storage; Environmental impact of energy conversion.

The participation to this conference represented a good way to inter-personal and inter-institutional communication, discussion and exchange of ideas based on relevant research themes specific to the thermodynamics field, and to identifying potential partners and collaborators.

NACOT2023 Organizing Committee
Viorel POPA
Florin POPESCU
Ion V. ION
Krisztina UZUNEANU
Cristian IOSIFESCU
Mugurel BURCIU
Gabriel-Bogdan CARP
Ionel OPREA
Mariana LUPCHIAN
Gelu COMAN
Marcel DRĂGAN
Simona PARASCHIV
Michael FRĂTIȚA
Robert Mădălin CHIVU

Content from this work may be used under the terms of the Creative Commons Attribution 3.0 licence. Any further distribution of this work must maintain attribution to the author(s) and the title of the work, journal citation and DOI.
Published under licence by IOP Publishing Ltd

IOP Conf. Series: Materials Science and Engineering 1290 (2023) 011002 doi:10.1088/1757-899X/1290/1/011002

Peer Review Statement

All papers published in this volume have been reviewed through processes administered by the Editors. Reviews were conducted by expert referees to the professional and scientific standards expected of a proceedings journal published by IOP Publishing.

- **Type of peer review:** Double Anonymous
- **Conference submission management system:** Morressier
- **Number of submissions received:** 42
- **Number of submissions sent for review:** 39
- **Number of submissions accepted:** 33
- **Acceptance Rate (Submissions Accepted / Submissions Received × 100):** 78.6
- **Average number of reviews per paper:** 2.09375
- **Total number of reviewers involved:** 21
- **Contact person for queries:**
  Name: Ion V. Ion
  Email: ion.ion@ugal.ro
  Affiliation: "Dunarea de Jos" University of Galati - Thermal Systems and Automotive

Content from this work may be used under the terms of the Creative Commons Attribution 3.0 licence. Any further distribution of this work must maintain attribution to the author(s) and the title of the work, journal citation and DOI.
Published under licence by IOP Publishing Ltd

Methods for increasing the energy efficiency of fan systems using wide-angle plane diffusers with splitter vanes

Kliment Klimentov[1,2], Gencho Popov[1], Boris Kostov[1] and Lachezar Kamenov[1]

[1] Departmentof Heat, Hydraulics and Environmental Engineering, University of Ruse, Ruse 7017, Bulgaria

[2] kklimentov@uni-ruse.bg

Abstract. Fan systems used to transport fluids are widely used and have become a major consumer of electrical energy worldwide. A significant part of the available energy of transported air in HVAC systems has usually been lost (in the form of dynamic pressure) at the system outlet. It is well-known that using outlet diffusers leads to this energy loss be reduced. In some cases, it is necessary to use shortened wide-angle diffusers, which have a higher coefficient of minor energy loss (local resistance). In the present work, a numerical study of the possibilities of increasing the energy efficiency of wide-angle plane diffusers, using splitter vanes, is accomplished. Diffusers with expansion angles of 30, 45 and 60 deg studied. The results obtained indicate the impact of the number of splitter vanes on the minor energy (local resistance) loss coefficient of the studied diffusers.

1. Introduction

It is well-known that the use of a diffuser installed at the outlet of a fan system may lead to a reduction in energy consumption per unit volume of air (gas) transported in the system. Most often, to evaluate the energy efficiency of this type of system, the Specific Fan Power – SFP, defined in [1] is used. It can be determined by applying the following equation:

$$SFP = \frac{p}{\eta}, \ J/m^3, \tag{1}$$

where p is a fan total pressure, while η is its coefficient of efficiency.

Studies concerning the determination of the impact of the geometric features of conical diffusers on reducing the consumption of fan systems are presented in [2] and [3]. Often, air ducts, used in ventilation systems, are designed with a rectangular cross-section. This requires the use of flat or pyramidal diffusers having a rectangular cross-section. To assess the energy efficiency when using a diffuser in fan systems, it is necessary to know its minor energy loss coefficient $\zeta_{d,out}$. The values of $\zeta_{d,out}$ when installing the outlet of the system on flat diffusers with a rectangular cross-section are presented in [4]. Selecting a diffuser having a low value minor energy loss coefficient often leads to an excessive increase in the diffuser axial geometric sizes. On the other hand, the shortened diffusers, with higher expansion angles α, have higher values of $\zeta_{d,out}$, that reach $\zeta_{d,out} \approx 1$.

There are many modern studies on the flows in plane diffusers and on the methods used to reduce the flow separation, as well as to increase the static pressure recovery coefficient C_p, which is usually

Content from this work may be used under the terms of the Creative Commons Attribution 3.0 licence. Any further distribution of this work must maintain attribution to the author(s) and the title of the work, journal citation and DOI.
Published under licence by IOP Publishing Ltd

defined as the ratio of the actual static pressure rise $\Delta p = p_i - p_0$ in the diffuser to the mean inlet dynamic pressure [5]:

$$C_p = \frac{2\Delta p}{\rho \bar{U}_0} = \frac{2(p_i - p_o)}{\rho \bar{U}_0}, \tag{2}$$

where p_i is the static pressure in the considered (current) section of the certain diffuser, p_0 - the static pressure at the diffuser inlet; $\bar{U} = Q/S_0$ - average velocity at the diffuser inlet; Q - fluid volume flow rate; S_0 - the cross-section area at the diffuser inlet.

Research performed in this field, previously specified, can be divided into purely experimental, theoretical and those in which there are experimental and theoretical results combined. Results obtained after performing experimental studies of vaneless diffusers are published in [6], [7], [8] and [9].

In [10], [11], [12] and [13] numerical studies of vaneless diffusers are presented, and the results are compared with experimental ones.

Results obtained by performing numerical studies in the field of plane diffusers are presented in [14], [15], [16] and [17]. These papers are mainly focused on studying different methods used to improve the static pressure recovery coefficient.

In [5] the authors present the results obtained by performing an experimental study of two-dimensional diffusers with flat vanes and divergence angles of 30, 40, 60 and 80 degrees. The studied diffusers have and expansion rate of 4 and are being installed between rectangular cross-section pipes. On the basis of the experimental results obtained, the authors (in [5]) propose optimal combinations between the number of vanes, length and angle of vanes installation, in which the lowest coefficient of resistance is obtained.

In [18], the authors perform experimental studying concerning the determination of the impact of Reynolds number (Re) on the separation zone in a diffuser with a fixed angle (28 degrees) in two cases - having or not a splitter vane. The experiments are accomplished using water at Reynolds numbers Re=1000, 6000 and 12000. A similar study, but numerical, is presented in [19]. The studied diffuser has an expansion angle of 28 degrees, expansion rate 4.9, working fluid - dry air, ideal gas with a temperature of 20°C and a regime of movement, where Re=240000. Results obtained concerning the determination of the impact of the split vane on the distribution of the flow parameters and pressure recovery coefficient C_p, are presented.

In [20], the authors perform a theoretical study of flat and conical diffusers proving that, for the case of theoretically justified conditions, that it is possible to ensure a flow without separation at angles up to 20 degrees and static pressure-recovery coefficient, which corresponds to a diffuser with an angle of 7 degrees.

An interesting non-traditional approach to improve the efficiency of asymmetric plane diffusers is proposed by the authors of [21]. In this paper, a numerical study of an asymmetric plane diffuser with an angle of 25 degrees, when the flow has Re=9000 is accomplished. Implicit large-eddy simulation is employed to simulate the flow in the diffuser. An unconventional flow control method, namely, a cylindrical Karman-vortex generator (KVG) with different sizes and locations that induces periodic spanwise vortex shedding, is set upstream of the throat to suppress the flow separation [21]. The static pressure recovery coefficient C_p is increased by about 50%.

In [4] the author presents the most commonly applied methods used to reduce $\zeta_{d,out}$ of shortend wide-angle diffusers. As it was clearly described in [5], one of the proposed methods is based on dividing the diffuser with longitudinal dividing walls, thereby obtaining several diffusers with a smaller expansion angle β (figure 1).

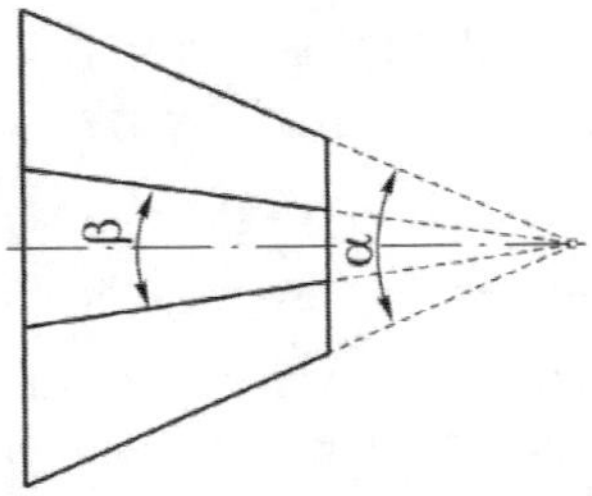

Figure 1. Specific angles of a wide-angle diffuser with dividing planes.

In [4] there is no data on the values of $\zeta_{d,out}$, concerning shortened diffusers with separating planes, and the study presented in [5] refers to a diffuser located between pipes. In the present work, a numerical study of diffusers installed at the system outlet having expansion angles of 30, 45 and 60 degrees is performed. Results related to the determination of the impact of the number and splitter vanes on the lminor energy loss coefficient of the studied diffusers and on improving the energy efficiency of fan systems when operating with these diffusers, are obtained.

2. Modeling the flow in plane diffusers without splitter vanes

2.1. Selection of numerical modeling method and turbulence model.

In the present section, a numerical study of the flow in plane diffusers having no splitter vanes is performed for three different wide-angle flat diffusers. Analyzing the reviewed literature sources makes it clear that different approaches are used to model turbulent separating internal flows: Direct Numerical Simulation (DNS) in [15], Large-Eddy Simulation in [14] и [16], Shear Stress Transport (SST) model in [19], Baseline Reynolds Stress Model (BSL RSM) in [11]. It is known that DNS and LES are modeling methods that require significant time and computational resources, and for thios reason two-equation models based on turbulent viscosity are most often applied to solve engineering problems - $k-\varepsilon$, $k-\omega$ and SST, or Reynolds Stress Models. A detailed comparative analysis of the applicability of these models in the modeling of plane assymetric diffusers is accomplished in [17] and it stands that RSM leads to the obtaining of most accurate results, despite the greater time for performing the calculations, compared to SST, $k-\varepsilon$, $k-\omega$ and the Spalart-Allmaras one-equation mode. In the present work, the Speziale-Sarkar-Gatski RSM (SSG RSM) model is selected for modeling the flow in the studied diffusers [22]. The calculations are performed by using ANSYS CFX 14.0.

2.2. Geometric sizes and computational meshes.

A diagram of the geometric models of the studied diffusers is presented in figure 2.

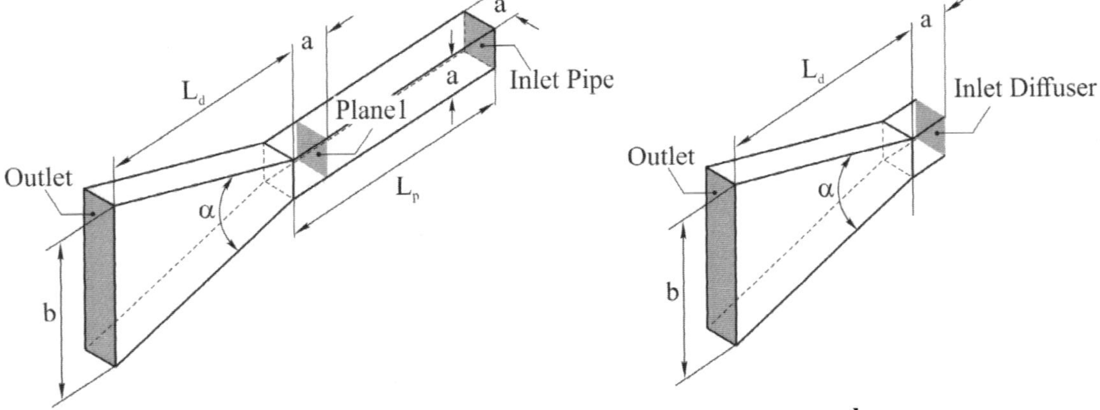

a **b**

Figure 2. Geometric sizes referred to the selected computational methods.

To determine a diffuser minor energy loss coefficient $\zeta_{d,out}$, when being installed at the system outlet it is necessary to use a model, in which there is a straight section of length $L_p = 10D_h$ in front of the diffuser inlet (figure 2-a). The hydraulic diameter of a square section is $D_h = a$. To reduce the computation time, the flow in the straight length section L_p is modeled in case of having not a diffuser at the system outlet. The resulting velocity profiles, as well as those of turbulent eddy dissipation and turbulent kinetic energy in the Plane 1 section, are collected and then presented in special formats, which are finally applied at the inlet of the Inlet Diffuser – figure 2-b. The models sizes referred to figure 2 are presented in table 1.

Table 1. Geometric model sizes referred to the studied diffusers.

α (deg)	*a* (mm)	*b* (mm)	L_p (mm)	L_d (mm)	*Mesh1* *NN*	*Mesh2* *NN*	*Mesh3* *NN*
30	60	240	600	355.88	238702	355142	471582
45	60	240	600	217.28	238702	355142	471582
60	60	240	600	155.88	238702	355142	471582

Three meshes, labeled Mesh1, Mesh2, and Mesh3, respectively, are generated to study the vaneless planar diffusers. The meshes are structured, with hexahedral elements, and the number of nodes (NN) for each one of them are presented in table 1.

2.3. Computational procedure and results obtained.

The computation procedure is performed in ANSYS CFX. The following work fluid is selected to run the simulations – air with density $\rho = 1.284\ kg / m^3$ and viscosity $\mu = 1.725 \cdot 10^{-5}\ Pa \cdot s$. At the entrance to the straight section, a uniformly distributed velocity is set so that the average speed in Plane1 to be $\bar{U}_0 = 22.391\ m / s$. Moreover, the Reynolds number in this section is $Re_0 = 100000$. The turbulent model Speziale-Sarkar-Gatski Reynolds Stress Model (SSG RSM) [22], analysis type – transient, is used. For initial conditions of transient simulation, results obtained after steady state analysis are used. Timesteps for transient simulation is selected in the range $\Delta t \le 0.0001 s$, in which it is obtained that the Courant number $C = u\Delta t / \Delta h < 1$, where u represents the characteristic flow velocity, and Δh - typical cell size. Number of timesteps per run was set as 1000. Maximum iteration was set as 100. Simulation convergence was achieved when the residual error was lower than 10^{-3}, in which root mean square error is obtained as lower than 10^{-5}.

Table 2 contains the results of the calculations performed. The values of $\zeta_{d,out}$ are determined by applying the following equation [23]:

$$\zeta_{d,out} = 1 - \frac{2\Delta p}{\rho \bar{U}_0} = 1 - \frac{2(p_{out} - p_o)}{\rho \bar{U}_0}, \tag{3}$$

where p_{out} is the static pressure at the diffuser outlet. For validating the results obtained it is necessary to determine the relative error (deviation) Δ, as for this this aim the following equation is applied:

$$\Delta = \frac{\zeta_{Id} - \zeta_{CFX}}{\zeta_{Id}} 100\%, \tag{4}$$

where ζ_{Id} is the value of the diffuser minor energy loss coefficient, experimentally obtained and published in [4], and ζ_{CFX} - represent the values obtained after performing numerical calculations. Number of nodes for the three meshes are also given in the second row of the table. The values of ζ_{Id} at $Re_0 = 100000$ and a straight section installed in front of the diffuser with length $L_p \ge 10D_h$, are [4]:

$\zeta_{Id} = 0.72$ at $\alpha = 30\ deg$; $\zeta_{Id} = 0.88$ at $\alpha = 45\ deg$ and $\zeta_{Id} = 1.02$ at $\alpha = 60\ deg$.

Table 2. Results obtained after performing numerical simulations.

	Mesh1		*Mesh2*		*Mesh3*	
NN	238702		355142		471582	
α	ζ_{CFX}	Δ	ζ_{CFX}	Δ	ζ_{CFX}	Δ
deg	-	%	-	%	-	%
30	0.740	2.78	0.738	2.5	0.737	2.36
45	0.868	1.36	0.874	0.68	0.877	0.34
60	0.949	6.96	0.960	5.88	0.959	5.98

From the data in Table 2, it can be seen that the maximum error, for all the studied variants, does not exceed 6%, at $\alpha = 60\ deg$ and Mesh3. Furthermore, it is clear that when increasing the number of grid cells from 238702 to 471582, which is nearly two times higher, the relative change in the value of ζ_{CFX} is 0.41% at $\alpha = 30\ deg$, 1.03% - at $\alpha = 45\ deg$ and 1.04% at $\alpha = 60\ deg$. When increasing NN from 355142, at Mesh2, to 471582, at Mesh3, which is nearly 25%, the relative change in the value of ζ_{CFX} is 0.14% at $\alpha = 30\ deg$, 0.34% - at $\alpha = 45\ deg$ and 0.10% at $\alpha = 60\ deg$. This means that the results obtained with Mesh3 can be considered reliable and accurate enough for performing engineering calculations..

3. Modeling the flow in plane diffusers with splitter vanes

3.1. Geometric models and computational meshes.

A general view of a plane diffuser with splitter vanes is shown in figure 3. According to the recommendations in [4], the separating walls installed before the diffuser inlet must be parallel directed to its axis, and the length of the straight section is to be $l_0 \geq 0.1a$. In the present study it is assumed that $l_0 = 0.01\ m$.

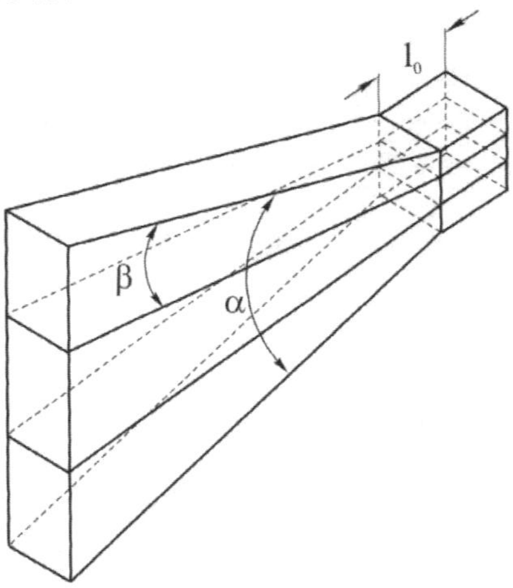

Figure 3. General view of a plane diffuser with splitter vanes

In the present study, the angle β between splitter vanes is assumed to be the same between all walls. In such a case, this angle is determined by applying the equation below:

$$\beta = \frac{\alpha}{z+1}, \quad (5)$$

where z is the number of splitter vanes.

In developing the geometric model, spliter vanes are assumed to have infinitely thin walls. In [4], the authors recommend that: at $\alpha = 30\ deg$, $z = 2$, and at $\alpha = 45$ and $60\ deg$, $z = 4$. In the present work, geometric models have been developed at $\alpha = 30 deg$ with $z = 2$ and $z = 3$. At $\alpha = 45\ deg$ and $\alpha = 60\ deg$, $z = 4$ and $z = 5$.

For each variant of a diffuser with splitter vanes, two meshes, labeled M1 and M2, with a different number of nodes are generated. NN values are given in Table 3.

Table 3. Number of Nodes (NN) referred to the previously generated computational meshes.

	z=2		*z=3*		*z=4*		*z=5*	
α	M1	M2	M1	M2	M1	M2	M1	M2
deg	-	-	-	-	-	-	-	-
30	380562	562602	385769	709177	-	-	-	-
45	-	-	-	-	390976	512336	578223	735458
60	-	-	-	-	390976	512336	578223	735458

3.2. Turbulence model and computation procedure.

When modeling the flow in the studied plane diffusers with split vanes once again the SSG RSM turbulence model is selected. The calculations are performed by using ANSYS CFX 14.0 under the conditions specified in subsection 2.3.

3.3. Results obtained after performing numerical simulations.

The values of the coefficient ζ_{CFX} are presented in table 4.

Table 4. Values of ζ_{CFX} , obtained after performing numerical simulations.

	z=2		*z=3*		*z=4*		*z=5*	
α	M1	M2	M1	M2	M1	M2	M1	M2
deg	-	-	-	-	-	-	-	-
30	0.623	0.619	0.559	0.557	-	-	-	-
45	-	-	-	-	0.665	0.670	0.611	0.613
60	-	-	-	-	0.784	0.784	0.726	0.719

The highest relative variation in the value of ζ_{CFX} is observed for the diffuser with $\alpha = 60\ deg$ and $z = 5$. When NN changes from 578223 to 735458 (by 21.4%), the relative variation of ζ_{CFX} is less than 1%.

In the analysis of the results presented later in the work, the results obtained with M2 meshers are going to be used.

4. Analysis of the results obtained

Figure 4 represents relationships of type $\zeta_{CFX} = f(z)$ concerning the three types of studied diffusers.

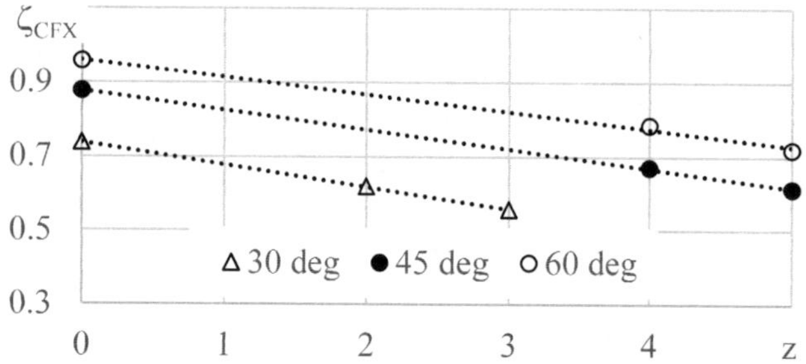

Figure 4. Relationships presenting the coefficient ζ_{CFX} as function of the number of splitter vanes z at different plane diffuser angles

The results obtained clearly indicates that when using splitter vanes it is possible to reduce the diffuser minor energy loss coefficient $\zeta_{d,out}$. At the diffuser with an angle 30 deg this coefficient $\zeta_{d,out}$ is reduced by 16.0% when using 2 splitter vanes and by 24.4% when using 3 splitter vanes. The greatest effect of splitter vanes is seen in the angled diffuser 45 deg. In this case the studied coefficient $\zeta_{d,out}$ is reduced by 23.6% when using 4 vanes and by 30.1% when using 5 vanes. At the diffuser with an angle 60 deg the approximate reduction of $\zeta_{d,out}$ is by 18.2% when using 4 vanes and by 25.0% when using 5 vanes. From the results obtained, it can be clearly seen that the greatest effect of the use of splitter vanes, concerning the studied configurations of diffusers, is at an angle of 45 deg and z=5 vanes.

For describing the studied range $\zeta_{d,out} = f(\alpha, z)$ the following equation is to be used:

$$\zeta_{d,out} = 0.5 + 0.007896\alpha - 0.05z, \tag{6}$$

where the coefficient of determination is $R^2 = 0.9831$.

As mentioned in the Introduction, the use of a diffuser installed at the outlet of a fan system can improve the system energy efficiency. In accordance with [3], the reduction of specific fan power ΔSFP when using a diffuser installed at the system output and achieving the initial operating mode after VFD flow regulation is applied, may be determined using the following equation:

$$\Delta SFP = 0{,}001\frac{p_1 - p_2}{\eta} = 0{,}001\frac{\Delta p}{\eta}\left[\frac{kW}{m^3/s}\right], \tag{7}$$

where η is the fan coefficient of efficiency. It is also stated in [3] that:

$$\Delta p = \rho\frac{Q^2}{2S_0^2}\left(1 - \zeta_{d,out}\right), \tag{8}$$

where Q is the system volume flow rate; S_0 - the diffuser inlet cross-section area. By taking equations (7) and (8) into account, it is clear that for the diffuser with $\alpha = 45\ deg$ ΔSFP decreases by 12% at $z = 0$ and by 38.7% at $z = 5$, which represents a deviation of 26.7%.

5. Conclusion

The paper presents the results obtained after performing numerical modeling of the flow in flat diffusers without and with splitter vanes. The minor energy loss coefficients of the studied diffusers $\zeta_{d,out}$ are also determined. For the diffusers without splitter vanes, the numerical values obtained are compared with experimental data taken from [4]. The relative difference in experimental and numerical values does not exceed 6% for the angle diffuser $\alpha = 60\ deg$, while in case of $\alpha = 30\ deg$ and $\alpha = 45\ deg$ these values are 2.36% and 0.34%, respectively. In diffusers with splitter vanes, a reduction of $\zeta_{d,out}$ is indicated, as the most significant relative decrease of 30.1% is referred to the case of a diffuser with $\alpha = 45\ deg$ and $z = 5$ splitter vanes.

Based on the numerical results obtained, a specific relationship of the type $\zeta_{d,out} = f(\alpha, z)$, which is valid for the studied ranges of the parameters α and z, is established.

The results obtained can be used to analyze and increase the energy efficiency of fan systems with a diffuser installed at the system outlet.

Acknowledgments

This current paper presents the results of working on project No. 2023 - AIF - 03, funded by the Research Fund of the University of Ruse.

References

[1] EN 13779: 2006 Ventilation for Non-Residential Buildings — Performance Requirements for Ventilation and Room–Conditioning System.

[2] Popov G, Klimentov K, Kostov B and Dimitrova R 2020 Increasing the energy efficiency of fan systems by using outlet diffuser *7th International Conference on Energy Efficiency and Agricultural Engineering (EE&AE)*, 4, pp 1-4, doi: 10.1051/e3sconf/202020704004

[3] Popov G, Klimentov K and Kostov B, Specific features in determining the energy efficiency when using a diffuser at the outlet of fan systems, Mechanics of machines 126, pp 64-68, 2022

[4] I. Idel'chik 2007 *Handbook of hydraulic resistance*. 4th revised and augmented edition. Moskow

[5] Yamazato E and Irabu K 1975 Performance of wide angle two-dimentional diffusers with vane systems Bulletin of Sciense&Engineering Division, University of Ryukus. Engineering (**8**): pp 1-15

[6] Salim B 2013 Effect of goemeterical parameters on the performance of wide angle diffusers *International Journal of Innovative Research in Science, Engineering and Technology* Vol. **2**, Issue **9**, pp 4178-4191

[7] Cherry E, Elkins C and Eaton J 2008 Geometric sensitivity of three-dimensional separated flows *International Journal of Heat and Fluid Flow* **29** pp 803–811

[8] Cherry E, Elkins C and Eaton J 2009 Pressure measurements in a three-dimensional separated diffuser *International Journal of Heat and Fluid Flow* **30** pp 1–2

[9] Sayles E and Eaton J 2014 Sensitivity of an asymmetric, three-dimensional diffuser to inlet condition perturbations *International Journal of Heat and Fluid Flow* **49** pp 100–107

[10] Tornblom O 2006 *Experimental and computational studies of turbulent separating internal flows* Doctoral thesys, Universitetsservice US AB, Stockholm

[11] Song X, Yu Z, Liu C et al. 2022 Calibration of RANS model constant based on data assimilation and accurate simulation of separated flow *AIP Advances* **12**, pp 095324-1-095324-9

[12] Obi S 1993 *Experimental and Computational Study of Turbulent Separating Flow in an Asymmetric Plane Diffuser* NINTH SYMPOSIUM ON "TURBULENT SHEAR FLOWS" Kyoto, Japan, August 16-1 8, 1993 pp 305-1-305-4

[13] Gullman-Strand J, Tornblom O, Lindgren B, Amberg G and Johansson A 2004 Numerical and experimental study of separated flow in a plane asymmetric diffuser *International Journal of Heat and Fluid Flow* **25** pp 451–460

[14] Tang H, Lei Y, Li X and Fu Y 2019 Large-Eddy Simulation of an Asymmetric Plane Diffuser: Comparison of Different Subgrid Scale Models *Symmetry 2019* **11**, pp 1-18

[15] Ohta T and Kajishima T 2010 Analiysis of non-steady separated turbulent flow in an asymmetric plane diffuser by Direct Nummerical Simulations Journal of fluid science and technology **5(3)** pp 515-527

[16] Shneyder H, von Terzi D, Bauer H J and Rodi W 2010 Reliable and Accurate Prediction of Three-Dimensional Separation in Asymmetric Diffusers Using Large-Eddy Simulation *Journal ofFluids Engineering* **132(3)** pp 1-7

[17] Berdanier R 2011 *Turbulent flow through an asymmetric plane diffuser* Technical Report Purdue University

[18] Miao Guo et al 2021 *Experimental influence of a splitter vane on the flow fields in a wide-angled diffuser* IOP Conf. Ser.: Earth Environ. Sci. 774 012051

[19] Meng X, Zuo Z, Nishi M and Liu S 2020 A Numerical Study on the Flow Mechanism of Performance Improvement of a Wide-Angle Diffuser by Inserting a Short Splitter Vane *Processes* **8**, 143 1-14 doi:10.3390/pr8020143

[20] Zaryannkin A, Cherkasov M and Lavyrev I 2021 Theoretical foundations of increasing the efficiency of wide-angle diffusers and practical results of new solutions to this problem *MATEC Web of Conferences* ***345**, 00035 Power System Engineering*

[21] Yang J, Zhang Y, Chen H and Fu S 2020 Unsteady flow control of a plane diffuser based on a Karman-vortex generator *AIP Advances* **10**, 055314 1-15

[22] Speziale, C G, Sarkar S and Gatski T B 1991 Modeling the Pressure-Strain Correlation of Turbulence: an Invariant Dynamical Systems Approach *Journal of Fluid Mechanics,* Vol. **227**, pp. 245-272.

[23] Miller D 1990 *Internal flow systems* BHRA, The Fluid Engineering Centre

Temperature during the milling process of aluminum alloys using contact and non-contact methods

E N Patru, M Bica, D Panduru and N Craciunoiu

Faculty of Mechanics, University of Craiova, Calea Bucuresti, 107, Craiova, Romania

nick_emyl@yahoo.com

Abstract. In this paper some experimental determinations on the temperature during the milling process of aluminum alloys (5083, 6082 and 7075) were conducted, by using two methods: with contact and non-contact, for the different feed speeds. For this reason, an adequate installation was used, so that the temperature can be measure simultaneously. The results are presented using graphical dependencies of the temperature as function of the feed speed, for each type experiment, and then a synthesis regarding the maximum values of the temperature as function of the material type and the value of the feed speeds is presented.

1. Introduction

The aluminum alloys are the most and common materials used in aerospace, automotive, food industries and not only, due to high specific strength, low cost, easy processing, etc. [1], [2].

It can be added that aluminum alloys have been extensively used for the manufacture of engineering components and structures where corrosion resistance and lightweight is paramount, [3].

The milling process of the aluminum alloys is one of the efficient machining operations, [4].

In order to monitoring the cutting process, the temperature can be considered an important factor with influence on the tool wear, dimensional and geometrical deviations, surface roughness etc.

The values of the temperature and the heat generation during milling process of aluminum alloys was the studied in many research papers, and here, some of these papers are selected.

In the synthesis works, [5] and [6] the authors presents, based on wide literature, aspects regarding machining and machinability behaviour of the aluminum alloys.

In [7] the influence of machining parameters on the heat generation in milling process of an aluminum alloy used in aerospace applications is analysed. The authors compare the influence of a chamfered cutting and sharp cutting edges on stability and heat generation for different feed per tooth and spindle speed.

In order to optimize the process parameters for milling process of 6063, aluminum alloy in [8] the authors used Taguchi L_{18} array for study the effect of variation for the number of flutes, the spindle speed, depth of cut and feed rate on the workpiece temperature.

Some aspects regarding the thermo mechanical process in the primary shear zone, when dry machining the aluminum alloy AA2024–T351, is presented in [9]. Also, for the aluminum alloy, thin plate, 2024 –T3, using specific location for thermocouples, in [10] is evaluated temperature of the workpiece subsurface during two milling operations, slotting and ramping, and as results the temperature increase with the feed reduction, the authors conclude.

Content from this work may be used under the terms of the Creative Commons Attribution 3.0 licence. Any further distribution of this work must maintain attribution to the author(s) and the title of the work, journal citation and DOI.
Published under licence by IOP Publishing Ltd

In [11] a practical approach to predict the cutting temperature in trochoidal milling was established using different ways, and finally, as the authors show, the cutter edge temperature patterns during trochoidal and side milling were extensively analyzed.

In the paper [12] a systematic thermodynamic analysis of the dry milling process, based on a data acquisition system, with different types of sensors, was used to determinate the temperature and the heat flow into the tool for different cutting velocity. They, also, compare the measured values with the simulated values for temperature.

An adequate position for the five thermocouples to measure the temperature during milling process is used in [13]. They specify that the accuracy of the results depends on the distance between the thermocouples and the active (cutting) surface of the workpiece. If the thermocouples are closer to active surface (0.5 mm in their study), the greater accuracy will be. In order to determinate the temperature, the experiments was conducted for different cutting speeds and feeds rate in the case of the moving point or plane heat source. The measured values of the temperatures were used to estimate the net heat flowing into the workpiece.

The influence of cutting strategy on the temperature and surface flatness in face milling process is experimentally determinate in [14] using six thermocouples positioned on the workpiece. Also the heat transfer was simulated numerically using finite element method and finite differences method, and then the results were compared.

This paper present the results, in terms of temperature, obtained for the milling process of three aluminum alloys, EN AW-5083, EN AW-6082 and EN AW-7075, after previously determination the probes emissivity. The temperature was measured by using six thermocouples (contact method) and an infrared portable thermometer (non-contact method).

For the first method, with contact, six thermocouples were installed along the workpiece, but in such way that the temperature is measured in the workpiece body.

All the six thermocouples are arranged in the holes, so that the temperature to be measured at the middle of the workpiece at 0.5 mm under the cutting surface.

For the second method, non-contact, an infrared thermometer, Optris MS Pro, was used. The thermometer was installed so that the distance between this and piece ensure a ratio distance/spot size 20/1, as it is recommended in the data sheet.

The milling temperature was simultaneous measured and the values are presented graphically in this paper.

It is important to mention that the main purpose of this paper is to analyze the difference, in term of the temperature, between the two methods.

2. Experimental setup

2.1 Materials and cutting tool

For the three types of aluminum alloys (EN AW-5083, EN AW-6082 and EN AW-7075) the workpieces were previously machining as is presented in figure 1.

In order to ensure a correct installation of the thermocouples, very important is to respect the dimensions (the distance between the holes, the depth of the holes) and the shape deviations (the surface flatness), for each faces of the workpiece,

The workpieces were previously machined so that the temperature can be measured according to the three values of the cutting parameters (depth of cut, rotational speed and feed speed). Because six thermocouples are installed along the workpiece, for the three values of the each cutting parameter, are necessary to be drilled 18 holes for each of the three faces. Six for the first pass of the tool, another six for the next pass and six for the last pass of the tool (figure 1c).

The object of this paper is to present the results just for values of temperature measured as function of the feed speed.

Previously to the temperature measurement, after the workpieces were milled, the flatness deviation was measured and the thermocouples were installed (figure 2).

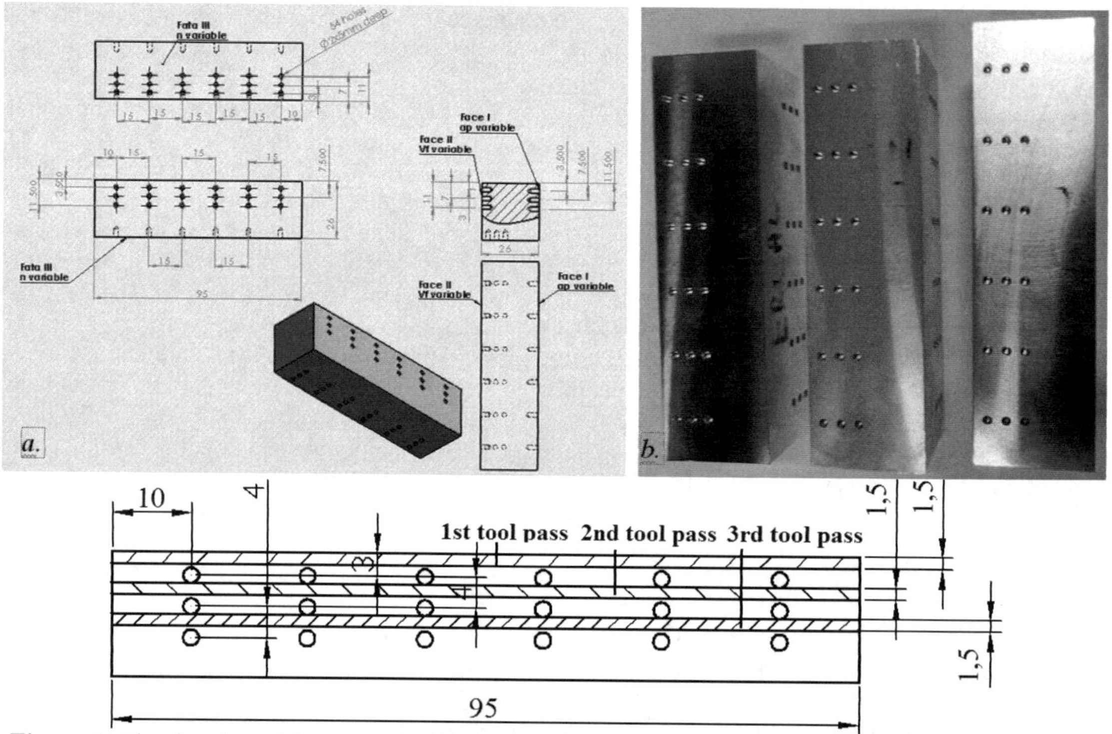

Figure 1. The drawing of the part with the faces and dimensions, *a.*, the workpieces prepared for the experiments, *b* and *c.* face I of the workpiece, with the three holes rows for temperature measurement by using different cutting speeds for the values of the depth of cut and feed speed constant

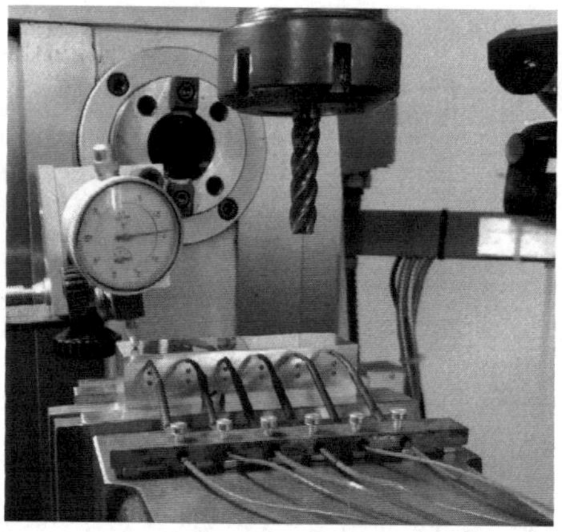

Figure 2. Test for the flatness control and positions of the thermocouples.

The holes were placed take in consideration the fact that for each face (I, II and III, see figure 1*a*), during the entire experimental determinations, it was used different parameters for the cutting regime (depth of cut, speed feed and rotational speed).

The mechanical properties of the aluminum alloys, are: EN AW-5083 (Rp0,2 = 302 MPa, Rm = 176 MPa and elongation, 20%), EN AW-7075 (Rp0,2 = 532 MPa, Rm = 594 MPa and elongation, 10%), and EN AW-6082 (Rp0,2 = 295 MPa, Rm = 240 MPa and elongation, 8%).

The cutting tool, a carbide end mill, with ϕ10 mm diameter and 4 flutes, is presented in figure 3.

Figure 3. Carbide end mill, with ϕ10 mm diameter and 4 flutes.

2.2 Cutting conditions and experimental assembly for the temperature measuring

In order to measure the temperature, for each type of aluminum alloys, three probes were previously milled, as is presented in paragraph 2.1 and figure 1.

It is important to mention that the workpieces, was prepared to use different depth of cut, different feed speed and different cutting speed.

In this paper the results are presented, only, for the different feed speeds, the other parameters remaining constant (depth of cut and cutting speed, at the values of 1.5 mm and 22.619 m/min, respectively.

The cutting conditions were according with the ranges of the feed speeds and the rotational speeds of the tool for the milling machine UWF 80 E.

In order to have the correct measuring values of the temperature during the cutting processes, in general, and milling process of aluminum alloys, in particular, is to know the emissivity. This one is the measure of an object's ability to emit infrared energy and the values are from 0 (shiny mirror) to 1.0 (blackbody) [15].

The emissivity depends on numerous factors such as the type of material, surface roughness, microstructure, temperature, wavelength, etc. [16], [17]. In their experimental work, done for Al 6082, in the range of 200^0 C to 50^0 C, the authors show that the emissivity value is in the range 0.09 to 0.24. The value of the emissivity decrease with the decrease of the temperature and increase with an increase in the surface roughness.

To establish the correct value of the emissivity is a very difficult problem. One way can be by using recommends values from the technical book of the infrared thermometer (common material emissivity table). Another method is to compare the results of the infrared camera with the measurement from a contact thermocouple, or by using the method of the emissivity coating. Coating material can be a layer of the blackbody paint or vinyl electrical tape. But the best way is to use the experimental determination. In this paper the value of 0.107 was used for emissivity, as results of the experimental determinations presented in detail in [18], figure 4. The experimental installation was used both for the determination of the emissivity of the pure titanium and the aluminum alloys.

For this the sample was preheated and two thermocouples were in contact with the sample. These two thermocouples are coupled with a thermometer and temperature can be read on the display. In the same time, the sample temperature was measured by using an infrared thermometer, Compact Connect type, as in figure 4.

For the temperature measurement, the experimental assembly is presented in figure 5, having the following components: a portable infrared thermometer, Compact Connect type [19] with the measuring temperature range (-32^{0}C – 760^{0}C), emissivity adjustable between (0.1-1.00), precision ±1% and ratio distance/spot size 20/1, and six K-type thermocouples. The entire assembly was installed on a milling machine with digital read out, UWF 80 E type.

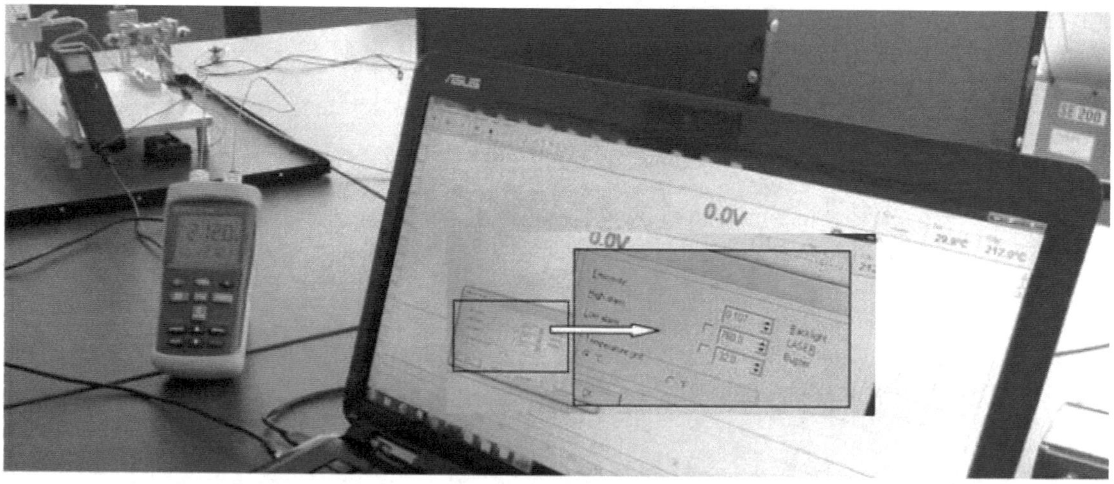

Figure 4. Part of the experimental determination of the emissivity.

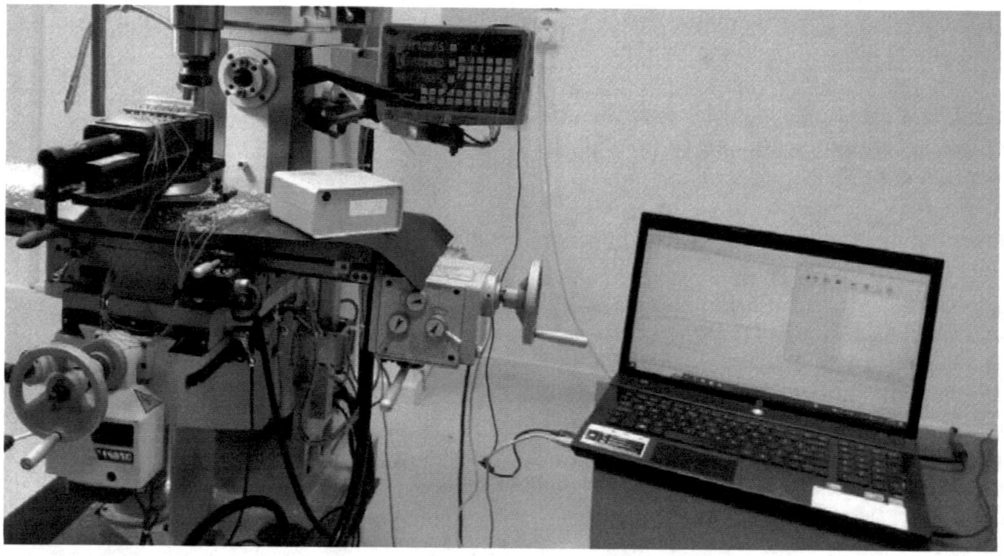

Figure 5. Experimental installation.

3. Results

3.1 Temperature results by using contact method with six thermocouples

The data were collected using the cutting conditions, portable infrared thermometer above described, and the six thermocouples connected with an acquisition data system to a personal computer.

For all data collected, using the ".dat" files, the graphical dependencies of the temperature as function of time, was obtained using Excel program. There are presented in figure 6 to figure 8, for the measuring of temperature with six thermocouples (the contact method) for the different feed speed in the case of 5083, 7075 and 6082 aluminum alloys, respectively.

3.2 Temperature results by using the non-contact method with infrared thermometer

For non-contact method, the values of the temperature are represented graphically in figure 9 to figure 11, for the different feed speeds in the case of 5083, 7075 and 6082 aluminum alloys, respectively.

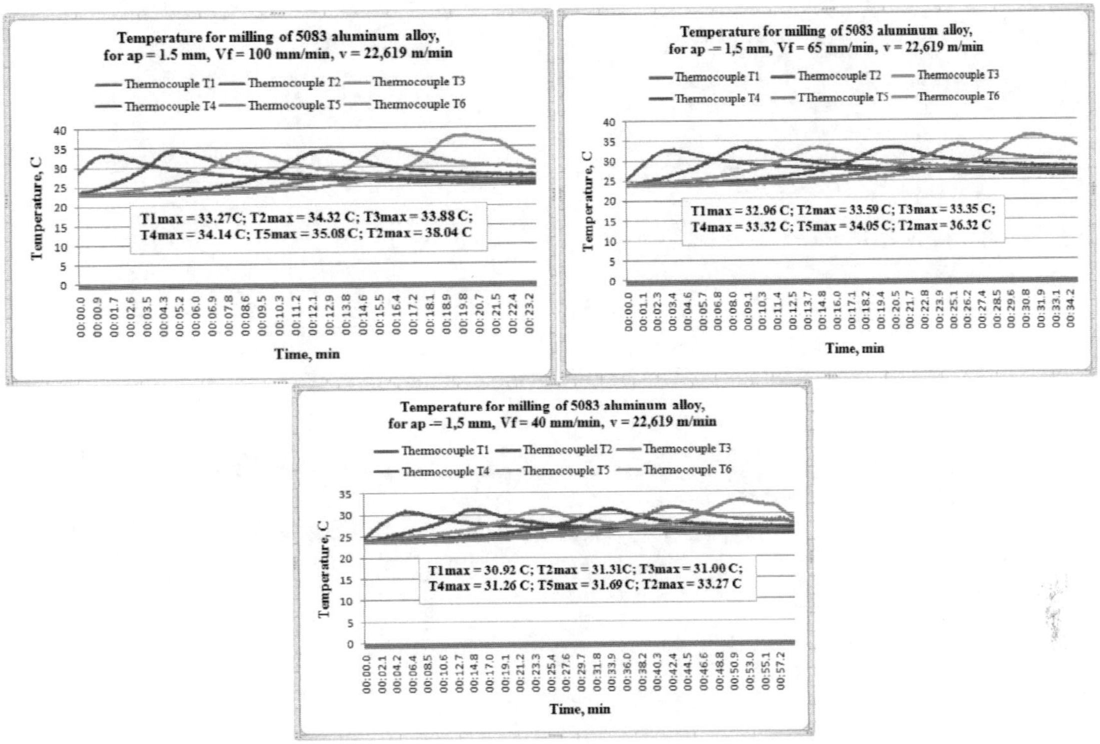

Figure 6. Temperatures for milling of 5083 aluminum alloy at different feed speeds.

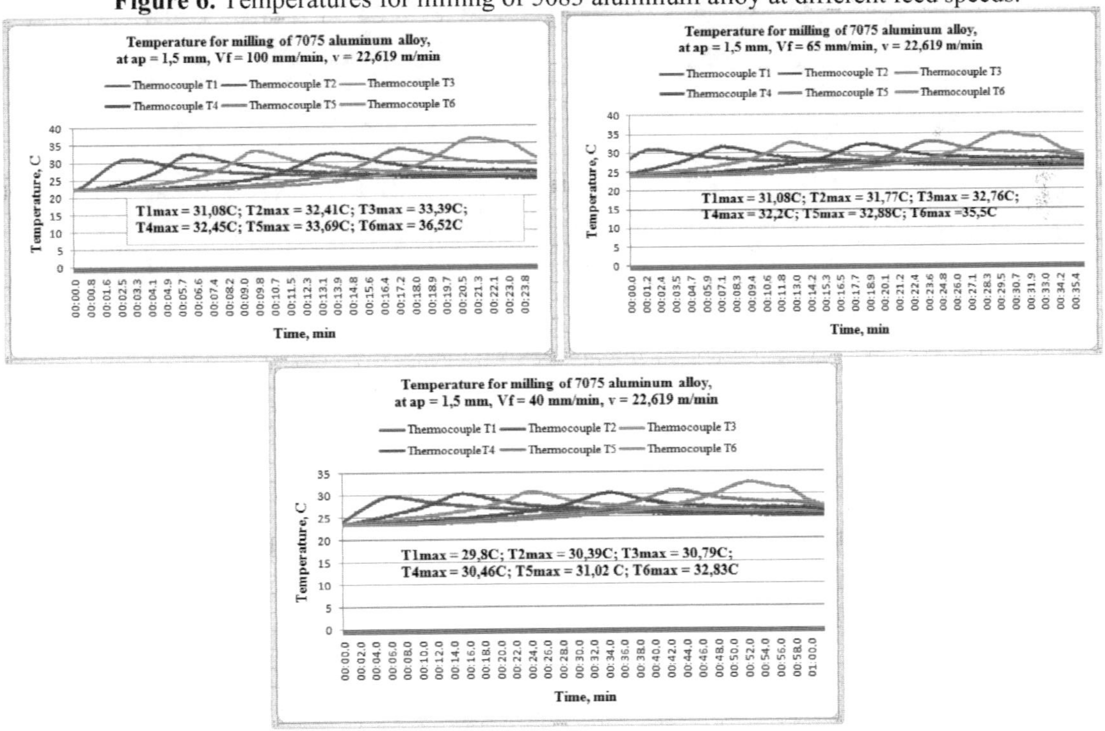

Figure 7. Temperatures for milling of 7075 aluminum alloy at different feed speeds.

IOP Conf. Series: Materials Science and Engineering 1290 (2023) 012002 doi:10.1088/1757-899X/1290/1/012002

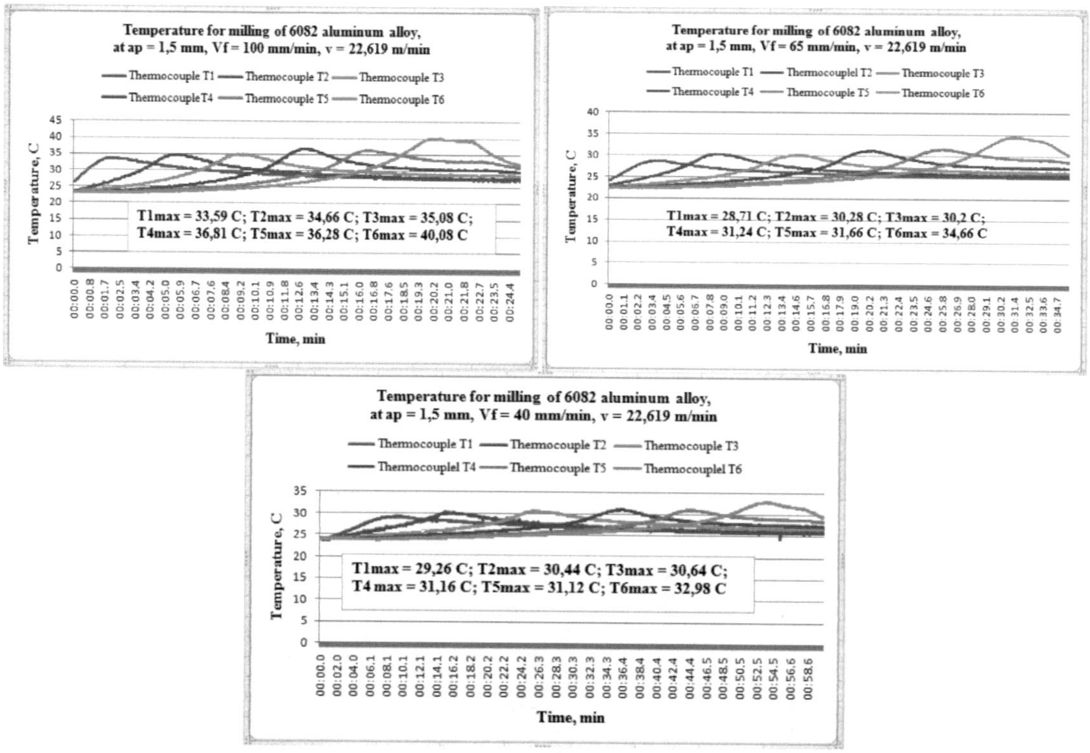

Figure 8. Temperatures for milling of 6082 aluminum alloy at different feed speeds.

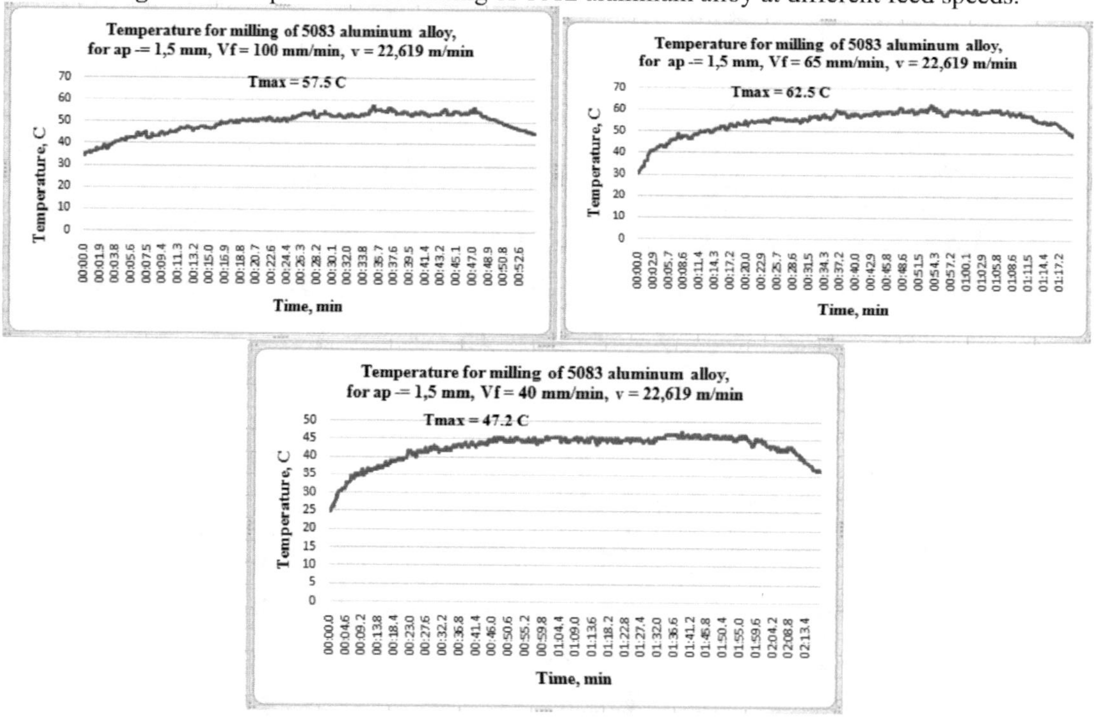

Figure 9. Temperatures for milling of 5083 aluminum alloy at different feed speeds.

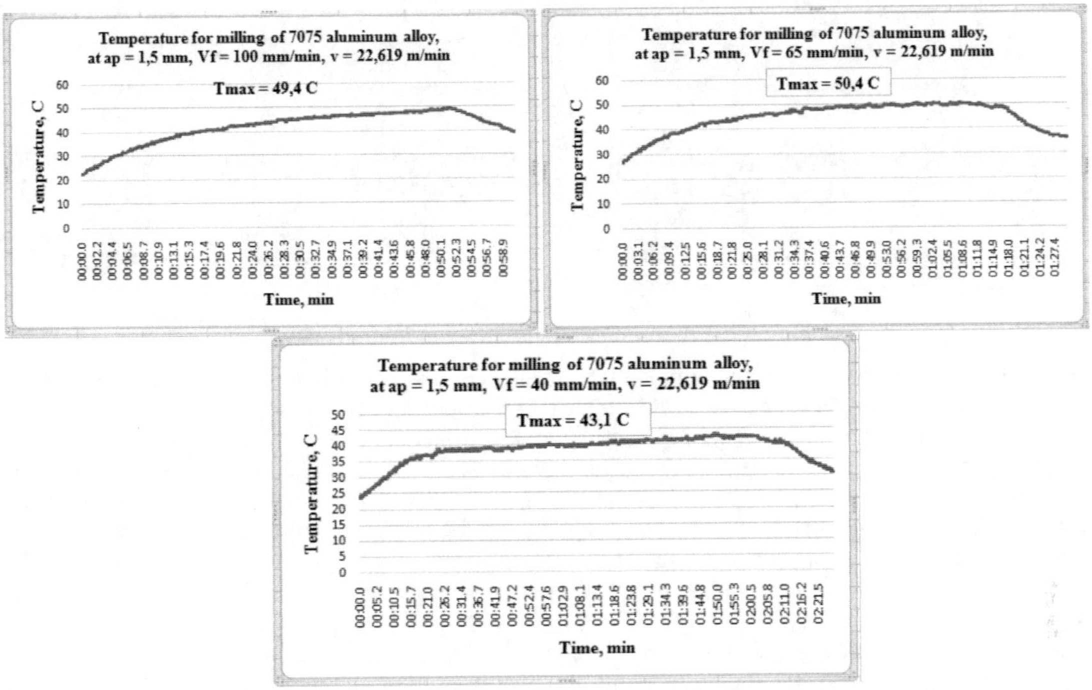

Figure 10. Temperatures for milling of 7075 aluminum alloy at different feed speeds.

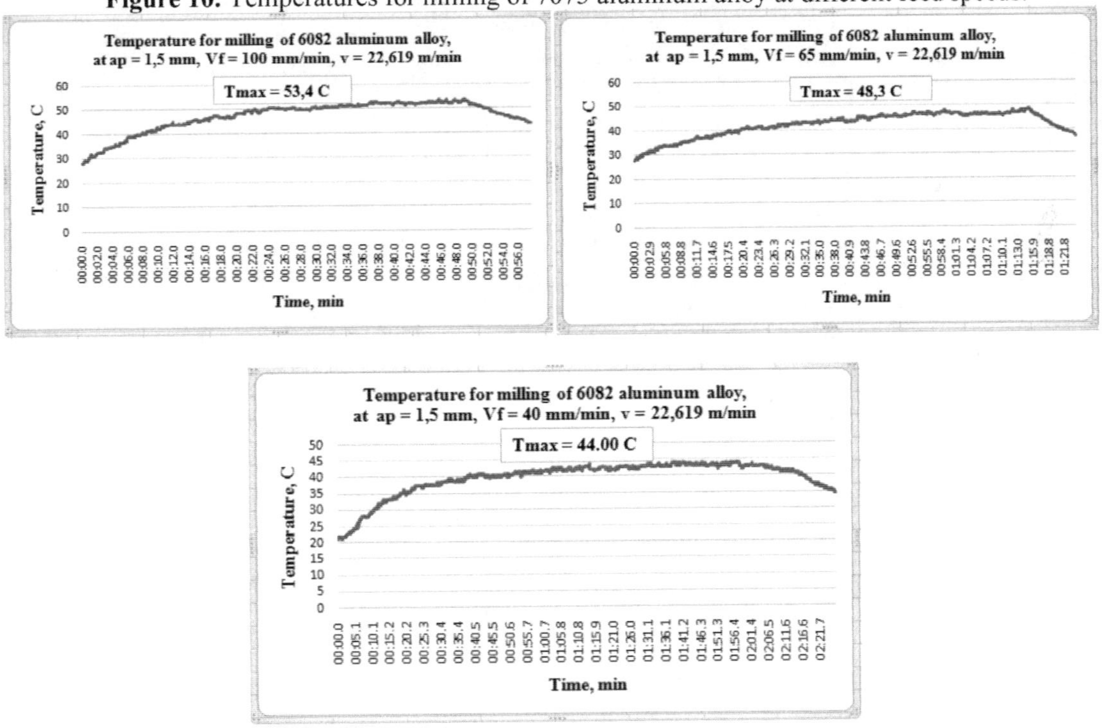

Figure 11. Temperatures for milling of 6082 aluminum alloy at different feed speeds.

By using the values of the measured temperatures, for the both methods, contact and non-contact, in figure 12 is presented a synthesis for the maximum values.

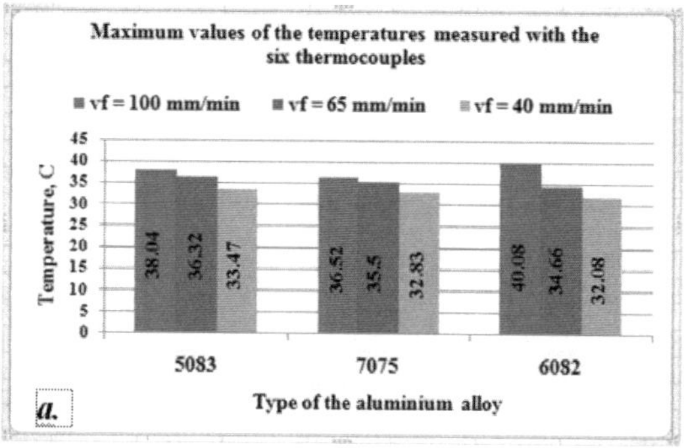

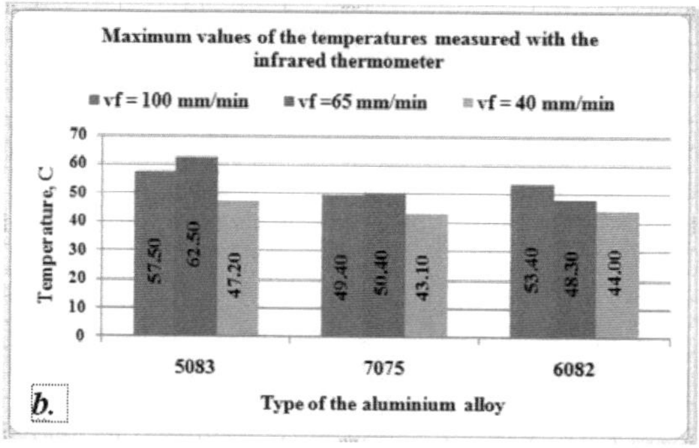

Figure 12. Maximum values of the temperatures for the three type of aluminum alloys at the different values of the speed feed for: *a.*contact method; *b.* non-contact method.

4. Conclusions

After the values of the temperature was measured, the representation from figure 6 to figure 8 was obtained for measuring with the six thermocouples and those of figure 9 to figure 11 was obtained for measuring with portable infrared thermometer.

Based on the graphical representation and the maximum values of the measured temperature by using the six thermocouples and infrared thermometer, in figure 12, *a* and *b,* a synthesis for the all three types of aluminum alloys and the three feed speed is represented.

From the experimental determinations and graphical representation it is observed the differences between measured values of the temperature for the contact and non-contact methods.

The higher values of the temperatures are obtained for the measurements with infrared thermometer (the non-contact method). That fact can be explained because, in this case, temperature is measured at the contact between tool and workpiece. For contact method temperature is measured at the middle of the workpiece and at 0.5 mm under the cutting surface.

From the point of view of the measured values, the maximum values of the temperature, as results of the measuring with the six thermocouples, are 40.08^{0}C for 6082 aluminum alloy, then 38.04^{0}C for 5083, both values for the feed speed 100 mm/min.

The minimum values are measured for the feed speed of 40 mm/min for all three type of aluminum alloys.

For the second method, non-contact, the maximum value, 62.50 ^{0}C is for 5083 and the feed speed 65 mm/min, then 57.50 ^{0}C, for the same material, but for 100 mm/min, the feed speed.

The lower value, 43.10^{0}C, was measured for 7075 aluminum alloy and 40 mm/min, the feed speed.

As a conclusion, the maximum values are measured for 5083 aluminum alloy, for the both methods and the feed speed at the value of 100 mm/min, with two exceptions: for 5082, the maximum values is 62.50 ^{0}C, for measuring with non-contact method and 65 mm/min, the feed speed, and also for 6082 and measuring with the contact method when the temperature, 40.08^{0}C, is the greater (for 100 mm/min, feed speed).

It is important to add that due to the difficulties of cutting process of aluminum alloy, especially built-up-edge phenomena, the temperature can, sometimes, be influenced, and this aspect will be the first step for the future experiments.

5. References

[1] Xianghui Huang, Jinyang Xu, Ming Chen and Fei Ren, 2020 Finite element modeling of high-speed milling 7050-T7451 alloys, *17th Global Conference on Sustainable Manufacturing,Procedia Manufacturing* **43** 471–478

[2] Yanzhong Zhang and Xiaohua Huang 2020 Simulation and research of aerospace material milling based on ABAQUS, *J. Phys.: Conf. Ser.* **1653** 012070

[3] Samuel A U et al 2021 Effect of Machining of Aluminium Alloys with Emphasis on

Aluminium 6061 Alloy – A Review *IOP Conf. Ser.: Mater. Sci. Eng.* **1107** 012157

[4] Zhiliaev A S and Kugultinov S D 2019 Mathematical simulation of thermal processes when milling aluminum alloy formed parts, *Journal of "Almaz – Antey" Air and Space Defence Corporation*, No. 2, 65-70

[5] Mário C. Santos Jr., Alisson R. Machado, Wisley F. Sales, Marcos A. S. Barrozo and Emmanuel O.Ezugwu 2016 Machining of aluminum alloys: a review *Int J Adv Manuf Technol* **86** 3067–3080

[6] Tusar Ranjan Soren, Ramanuj Kumar, Isham Panigrahi, Ashok Kumar Sahoo, Amlana Panda and Rabin Kumar Das 2019 Machinability behavior of Aluminum Alloys: A Brief Study, *ICMPC-2019, Materials Today: Proceedings* **18** 5069–5075

[7] Denkena B, Brüning J, Niederwestberg D and Grabowski R 2016 Influence of Machining Parameters on Heat Generation during Milling of Aluminum Alloys, *7th HPC – CIRP Conference on High Performance Cutting, Procedia CIRP* **46** 39 – 42.

[8] Bhirud N L and Gawande RR 2017 Optimization of process parameters during end milling and prediction of work piece temperature rise, *Archive of Mechanical Engineering*, VOL. **LXIV** Number 3, 327 – 346

[9] Atlati S, Moufki A, Nouari M and Haddag B 2017 Interaction between the local tribological conditions at the tool–chip interface and the thermomechanical process in the primary shear zone when dry machining the aluminum alloy AA2024–T351, *Tribology International* **105** 326–333

[10] Xavier Rimpault, Alexandre Il, Jean Francois Chatelain, Jean Francois Lalonde and Marek Balazinski 2018 Workpiece subsurface temperature study during aluminum skin milling in slotting and ramping. *8th CIRP Conference on High Performance Cutting (HPC 2018), Procedia CIRP* **77** 417-420

[11] Qi Deng, Rong Mo, Zezhong C. Chen and Zhiyong Chang 2020 An Analytical Approach to Cutter Edge Temperature Prediction in Milling and Its Application to Trochoidal Milling, *Appl. Sci.,* **10**, 1746; doi:10.3390/app10051746

[12] Augspurger T, Da Silva G, Schraknepper D, Mattfeld P and Bergs T 2020 Model-based monitoring of temperatures and heat flows in the milling cutter, The International Journal of Advanced Manufacturing Technology **107** 4231–4238

[13] Somayeh D. Farahani 2019 Estimation of distribution of moving heat source during milling of AISIH13 by inverse heat conduction method, *Journal of Theoretical and Applied Mechanics* Warsaw **57, 4** 1081-1093

[14] Lukasz Nowakowski, Michal Skrzyniarz, Slawomir Blasiak, and Marian Bartoszuk 2020 Influence of the cutting strategy on the temperature and surface flatness of the workpiece in face milling, *Materials*, **13**, 4542; doi:10.3390/ma13204542

[15] Soroush Masoudi, Mohammad Sima and Majid Tolouei-Rad 2018 Comparative study of ANN and ANFIS models for predicting temperature in machining, *Journal of Engineering Science and Technology* **13**, No. 1 211 - 225

[16] Lanc, Z, Štrbac, B, Zeljković M, Živković A and Hadžistević M 2018 Emissivity of aluminum alloy using infrared thermography technique, *MTAEC9, Materiali in tehnologije/Materials and technology* **52** 3, 323–327 doi:10.17222/mit.2017.152

[17] Lanc Z, Zeljković M, Štrbac B, Živković A, Drstvenšek I and Hadžistević M 2015 The determination of the emissivity of aluminum alloy AW 6082 using Infrared thermography, *Journal of Production Engineering, JPE* **18 (1)** 23-26.

[18] Panduru D, Bica M, Patru, E N and Craciunoiu N 2022 Simulation and experimental results on the temperature during turning process of commercially pure titanium, *Acta Technica Napocensis Series-Applied Mathematics Mechanics and Engineering*, **65**, 401-406, Special Issue, WOS:000911249300018

[19] Datasheet handheld thermometers of Optris MS series, https://www.optris.com

XXIIIrd National Conference on Thermodynamics with International Participation IOP Publishing
IOP Conf. Series: Materials Science and Engineering 1290 (2023) 012003 doi:10.1088/1757-899X/1290/1/012003

Study of hybrid nanofluids used in direct absorption solar collectors

G A Vardaru[1], G Huminic[1] and A Huminic[1]

[f1]Mechanical Engineering Department, Transilvania University of Brasov, Brasov, Romania

E-mail: gheorghe.vardaru@unitbv.ro

Abstract. The use of hybrid nanofluids in direct-absorption solar collectors is an efficient approach to harnessing solar energy. To address current energy challenges, in this study new hybrid nanoparticle aggregates containing distinct silver-based and iron-oxide nanophases (Ag+Fe2O3) were used to improve the photothermal conversion characteristics of direct absorption solar collectors. The main goal of the study was to investigate the effect of hybrid nanoparticle concentration and the efficiency of hybrid nanofluids on solar energy absorption performance. Experimental results showed that increasing the temperature of the nanofluid is higher than that of the base fluid, which means that nanoparticles dispersed in water have a stronger absorption of solar energy. The results also indicated that the efficiency decreases as increases time or temperature because the heat loss through the top surface of the sample increases at higher temperatures.

1. Introduction

The use of nanofluids in solar applications shows promising possibilities due to their exceptional heat transfer properties. The high thermal conductivity of dispersed nanoparticles allows efficient heat transport, rendering them ideal for cooling photovoltaic cells and harnessing solar energy [1].

The role of nanofluids in solar energy applications has rapidly gained attention. However, their advanced properties and practical application in thermal systems raise some concerns among scientists and engineers. A number of characteristics influence the effectiveness of nanofluids in improving the efficiency of solar thermal systems. These include thermal conductivity that improves heat exchange, density that allows optimal operating settings, pump efficiency, viscosity, specific heat that improves flow efficiency, and optical characteristics that facilitate the absorption of solar radiation [2]. Many scientific researchers has tested the potential of different nanofluids in various solar applications.

The concept of the direct absorption solar collector (DASC) was proposed by Minardi and Chuang in 1975.The main characteristic of DASC is to absorb solar radiation or energy through a heat transfer fluid (HTF). It is important to note that single-component nanoparticles were traditionally favored for DASC, despite their many drawbacks. For example, carbon nanotubes boast exceptional thermal conductivity and optical properties, but are relatively expensive and don't easily disperse uniformly.

On the other hand, ceramic oxides such as Al2O3, CuO, TiO2 and SiC are easier to disperse, but their thermal conductivities tend not to be as high. In addition, because their absorption spectra are found mainly in the visible region, their solar absorption capabilities are limited, with weak absorption in the near-infrared region.

To tackle these challenges, the researchers proposed a solution that supposes creating nanofluids using two or more types of nanoparticles. This revolutionary innovation has come to be known as 'hybrid

Content from this work may be used under the terms of the Creative Commons Attribution 3.0 licence. Any further distribution of this work must maintain attribution to the author(s) and the title of the work, journal citation and DOI.
Published under licence by IOP Publishing Ltd

nanofluids'. Various investigations demonstrated that hybrid nanofluids have the potential to efficiently harness solar radiation spectra, resulting in extended volumetric solar thermal absorption. This is achieved by exploiting the complementary optical absorption qualities of the various nanoparticles.
CuO-MWCNT hybrid nanofluid was prepared by Qu et al [3] to directly capture solar thermal energy. Their study showed that the spectral absorption properties of the hybrid nanofluids were significantly improved.
In a study by Tyagi et al. [4], the use of aluminum nanoparticles in water resulted in a 10% improvement in efficiency compared to traditional collectors.
Menbari et al. [5] focused on the optical properties and stability of CuO-Al2O3 hybrid nanofluids for use in direct absorption via parabolic trough solar collector (DASPTC).They found that increasing the nanoparticle loading and nanofluid flow rate can improve the thermal efficiency of DASPTC systems.
Meanwhile, Xuan et al [6] studied the photothermal conversion performance of nanofluids containing TiO2-Ag with core/shell structures and found that these hybrids absorb light over a wide range from UV to near-infrared. In addition, for photo-thermal conversion application, the CNT-SiO2/Ag hybrid nanofluid was prepared and tested with remarkable results. At a penetration distance of 1 cm, even low volume percentages of 0.005 gave a solar weighted absorption fraction of 74.5%.
In the preparation of water-based hybrid nanofluids, Chen et al. [7] used CuO and antimony doped tin oxide (ATO) nanoparticles. The tests showed that CuO-ATO nanofluids have strong light absorption in both the visible and near-infrared range and enhanced photo-thermal conversion capability, resulting in photo-thermal conversion efficiency of 92.5%.
In a study by Gupta et al. [8], the efficacy of aluminum oxide nanoparticles was evaluated in a range of volume fractions from 0.001 to 0.05. The results showed an impressive increase in efficiency of about 39.6% compared to the base fluid, water.
Valizade et al [9] performed spectrophotometric experiments on porous metal foams and water-based nanofluids, testing SiC and CuO materials.
In a study by Wang et al. [10], it was found that at a concentration of 20 ppm, MXene nanofluids exhibited 4.34% better photothermal conversion efficiency than graphene-based nanofluids. While the use of nanoparticles has been shown to improve solar thermal efficiency in DASC systems, the high cost, complex manufacturing processes, and potential toxicity of many nanomaterials have limited their full potential. As such, it is important to research and develop affordable, readily available and environmentally friendly nanofluids [11] with similar or better photothermal conversion effects.
In their research, Karami et al. [12] explored the potential applications of MWCNT nanofluids in volumetric solar collectors. Their findings revealed a significant improvement in the extinction coefficient of water-based MWCNT nanofluids, even at low particle concentration. These findings led to the expectation that the combination of these materials could produce distinct and remarkable characteristics.
Our current study focuses on evaluating the photo-thermal conversion efficiency of water and Ag-Fe2O3/water hybrid nanofluid for three different mass concentrations (0.50%, 0.75% and 1.0%) in a solar collector directly absorbing sunlight. The analysis is performed over a time interval of 5 to 65 minutes.

2. Experimental method

In order to determine the efficiency of hybrid nanofluids used in direct absorption solar collectors, a reflective solar simulator with a minimum lamp power of 1000 W, class AAA, was used. The simulator is ASTM E927-10 certified, with uniformity and temporal instability < 2%. The hybrid nanofluid sample was placed in the center of the solar simulator and covered with thin glass to minimize sample losses. Two K-type thermocouples were used to measure the temperature of the studied samples. The hybrid nanofluid samples were subjected to W/m2 radiation from the solar simulator for 3900 s, during which time the temperature was recorded. Solar radiation was measured using the Fluke IRR1-SOL instrument. Hybrid nanofluid samples were thermally insulated with glass wool to minimize heat loss. In this way,

IOP Conf. Series: Materials Science and Engineering 1290 (2023) 012003 doi:10.1088/1757-899X/1290/1/012003

major heat loss occurs only at the top surface of the sample, through convection and radiation combined in the ambient environment.

Figure 1. Solar simulator 1000 W

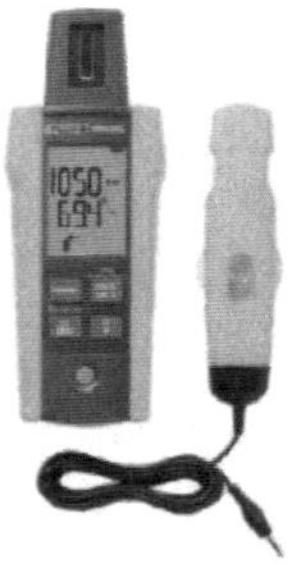

Figure 2. Fluke IRR1-SOL

3. Results and discussions

In order to determine the photothermal conversion capacity of the considered hybrid nanofluids, the efficacy was evaluated using equation (1).

$$\eta = \frac{m * c_p * \Delta T}{I * A * \Delta t} \quad (1)$$

where: m and c_p are the mass and specific heat of the studied fluid, ΔT - the temperature difference between the initial temperature and the instantaneous temperature, I - solar radiation, A - surface area, Δt - time frame

The effectiveness of hybrid nanofluids in solar thermal applications largely depends on the dependence of temperature variation on irradiation time.

In order to determine the temperature difference of Ag-Fe2O3 hybrid nanofluids as a function of irradiation time, the temperature was measured for water+4g/l CMCNa (carboxymethyl cellulose sodium) and three mass concentrations (0.50%, 0.75% and 1.0%) in the time interval 5-65 min.

The variation of temperature difference (T_m-T_0) with irradiation time is illustrated in **Figure 3.**

In **Figure 3.** one can notice that the addition of Ag-Fe2O3 nanoparticles significantly improves the photo-thermal performance of water.

The temperature of all fluids increased with time, but the maximum equilibrium temperature and temperature rise rate of the hybrid nanofluids surpassed that of water.

In particular, the temperature difference of hybrid nanofluids with a concentration of 0.5% reached a maximum of 17.35 °C after 65 minutes of irradiation. This rate was 30.84% faster than that of the base fluid.

The addition of a small amount of hybrid nanoparticles to the base fluid can significantly improve the photo-thermal conversion performance, as shown by the maximum equilibrium temperature difference which at a concentration of 1.0 % reached a maximum of 18.55 °C with approximately 35.31 % higher than that of the base fluid with the same irradiation time.

This trend is similar to that observed in other studies involving water-based hybrid nanofluids. In fact, Ag-Fe2O3 hybrid nanofluids exhibited even better photothermal properties due to their improved stability.

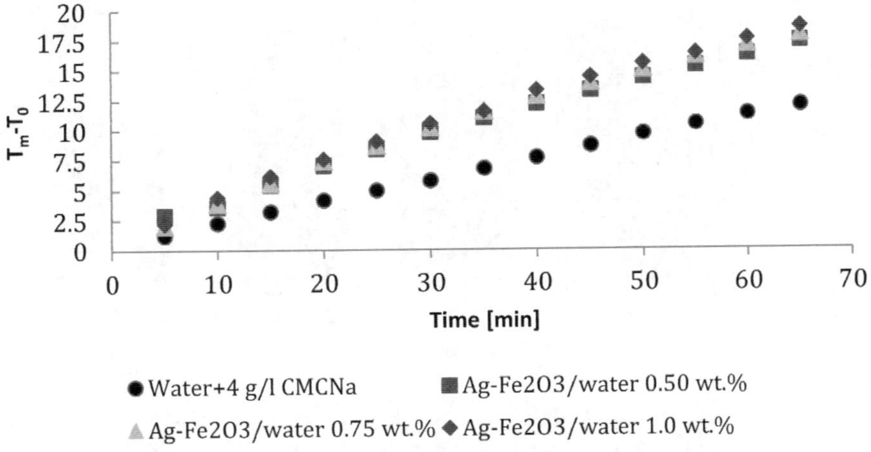

Figure 3. Temperature difference of Ag-Fe2O3 nanofluids as a function of irradiation time

Figure 4. shows the photothermal conversion efficiency of the working fluids, which was calculated to measure their performance. One can notice that in **Figure 4.**, the solar thermal efficiency of all fluids experiences a decrease attributed to increased heat loss to the environment as the fluid temperature increases. Hybrid nanofluids are particularly notable for their enhanced photothermal efficiency due to Ag and Fe2O3 nanoparticles, which possess enhanced photo-absorption capacity, exceeding that of the base fluid.

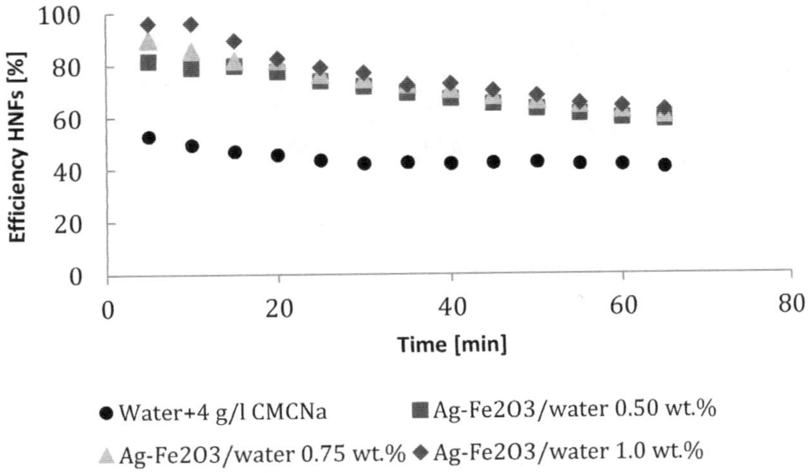

Figure 4. Efficiency of Ag-Fe2O3 nanofluids as a function of irradiation time

In **Figure 5** one can notice that efficiency decreases with increasing temperature, because heat loss through the top surface of the sample increases at higher temperatures.

IOP Conf. Series: Materials Science and Engineering 1290 (2023) 012003 doi:10.1088/1757-899X/1290/1/012003

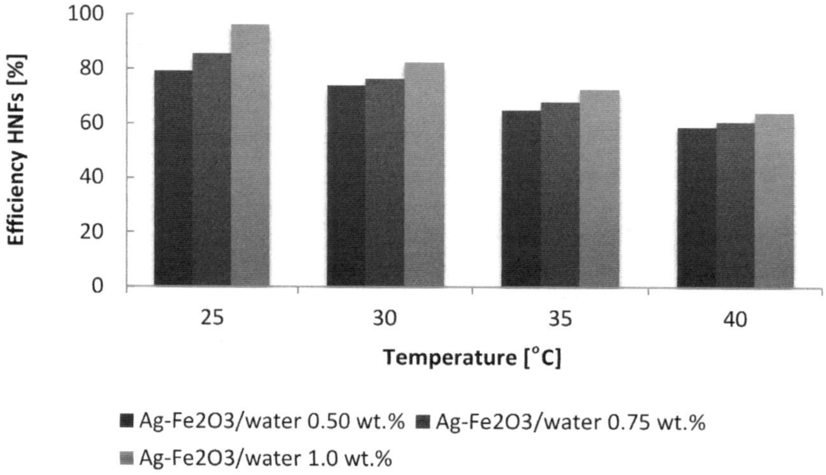

Figure 5 Variation of efficiency with temperature

Figure 6. illustrates the temperature increase at different nanoparticle concentrations in contrast to water (0%). All hybrid nanofluids caused higher temperature increases than water. The temperature increases ranged from 33.05 to 40.95 °C, thus the largest temperature increase exceeded that of the base fluid by 7.9 °C. This result demonstrates a significant improvement in the photothermal conversion capacity of the working fluids achieved by the inclusion of Ag and Fe2O3 nanoparticles.

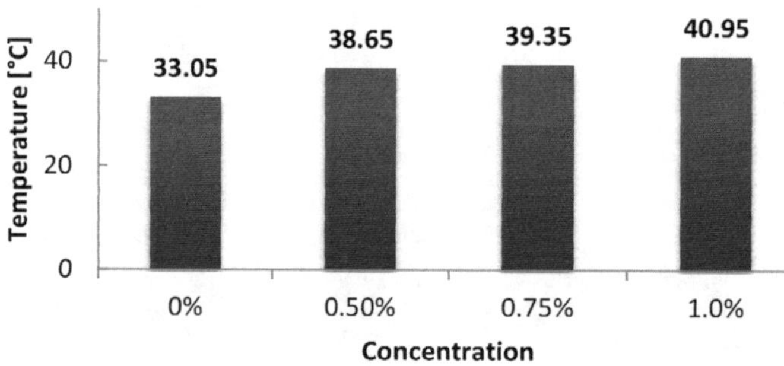

Figure 6. Temperature increase at different mass concentrations

The thermal efficiency of water and different concentrations of Ag-Fe2O3 nanoparticles is shown in **Figure 7**. The results show that the general thermal efficiency improves with increasing nanoparticle concentration. Specifically, the highest thermal efficiency of 96.05% is achieved with a mass fraction of 1%. Similarly, photothermal efficiency also increases with concentration, resulting in higher temperatures.

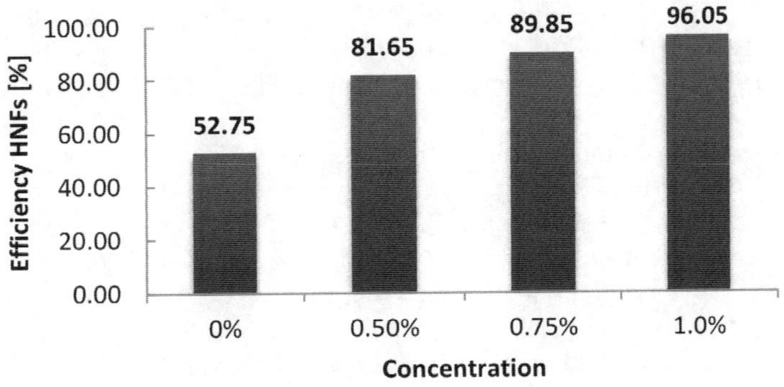

Figure 7. Photothermal conversion efficiency at different concentrations

Based on the results one can notice that the maximum efficiency was at 5 minutes of solar radiation after which it decreased with increasing temperature for all hybrid nanofluids studied. In this study, the highest efficiency was 96.05% at a concentration of 1 wt% per 5 minutes. However, these values decreased to 62.45% after 65 minutes of exposure. The mechanism behind this appearance can be outlined as follows: the incident solar radiation is fully assimilated within a certain period of time. As a consequence, nanofluids reach their maximum equilibrium temperature, which in turn is attributed to the increased adsorption energy. The surface temperature is consequently elevated due to this increased energy absorption [13]. The Ag-Fe2O3 hybrid nanofluids synthesized in this study show tremendous potential in DASC applications.

4. Conclusions

Measurements to determine the photo-thermal conversion efficiency were performed for water and Ag-Fe2O3/water hybrid nanofluid at three mass concentrations (0.50%, 0.75% and 1.0%) in the time interval 5-65 min.

Current research results on the use of Ag-Fe2O3 hybrid nanofluids in direct absorption solar collectors can be summarized as follows:

- There was a temperature rise of 40.95 °C, which exceeded the base fluid temperature by 7.9 °C.
- As the mass concentration increased, the hybrid nanofluids showed higher photo-thermal conversion efficiency. The maximum photo-thermal conversion efficiency of 96.05% was obtained for the Ag-Fe2O3 hybrid nanofluid with a concentration of 1 wt% in five minutes, which was 43.3% higher than that of water.
- The increase of temperature for the nanofluid is higher than for the base fluid, which means that nanoparticles dispersed in water have a stronger absorption of solar rays. After 65 minutes of irradiation, the Ag-Fe2O3 hybrid nanofluid with a concentration of 1% reached the highest temperature difference of 18.55°C. This represents an increase of 35.31% over the base fluid.

5. References

[1] Abderrahmane A, Hatami M, Medebber M A, Haroun S, Ahmed S E and Mohammed S 2022 *Alex. Eng. J.* **61** 5441

[2] Abderrahmane A, Naef A A Q, Abed M, Houssem L, Obai Y, Kamel G and Anas A 2023 *Appl. Therm. Eng.* **220** 119663

[3] Qu J, Zhang R, Wang Z and Wang Q 2019 *Appl. Therm. Eng.* **147** 390

[4] Tyagi H, Phelan P and Prasher R 2009 *J. Sol. Energy Eng. Trans. ASME.* **131** 0410041

[5] Menbari A, Alemrajabi A A and Rezaei A 2017 *Exp. Therm. Fluid Sci.* **80** 218
[6] Xuan Y, Duan H, Li Q 2014 *RSC Adv.* **4** 16206
[7] Chen N, Ma H, Li Y, Cheng J, Zhang C, Wu D and Zhu H 2017 *Sol. Energy Mater. Sol. Cells* **162** 83
[8] Gupta H K, Agrawal G D and Mathur J 2015 *Sol. Energy* **118** 390
[9] Valizade M, Heyhat M M and Maerefat M 2019 *Sol. Energy Mater. Sol. Cells* **195** 71
[10] Wang H, Li X, Luo B,Wei K and Zeng 2021 *Energy* **227** 120483
[11] Kumar L H, Kazi S N, Masjuki H H and Zubir M N M 2022 *Chem. Eng. J.* **429** 132321
[12] Karami M, Bahabadi M A A, Delfani S and Ghozatloo A 2014 *Sol. Energy Mater. Sol. Cells* **121** 114
[13] Chen M, He Y, Zhu J and Wen D 2016 *Appl. Energy* **181** 65

Acknowledgments:
This work was supported by grants from the Romanian Ministry of Education and Research, CNCS-UEFISCDI, projects number PN-III-P4-ID-PCE-2020-0353.

IOP Conf. Series: Materials Science and Engineering 1290 (2023) 012004 doi:10.1088/1757-899X/1290/1/012004

Development of minimalistic propulsion system for university nanosatellite

V S Baranova[1,a)], S V Liashkevich[1,a)] and V A Saetchnikov[1]

[1]Department of Physics and Aerospace technology, Belarusian State University, Minsk, Belarus

[a)]**Authors to whom correspondence should be addressed:** vbaranova@bsu.by and LeshkevichS@bsu.by

Abstract. The paper presents the results of development of a neutralizer-free high-frequency ion thruster for university nanosatellite propulsion systems which allow it to perform orbit correction tasks. The application of high-frequency ion thruster technologies allows the self-neutralization mode and is relevant for the development of engines for CubeSat format small satellites where the size and weight of the propulsion system must be as minimal as possible. A pair of shapes of neutralizer-free high-frequency ion thrusters with different accelerating region geometries are proposed. The analysis of plasma flow and the charged particles acceleration was carried out. The trajectory and acceleration of a quasi-neutral plasma flow was calculated for crossed electric and magnetic fields. The cyclotron resonance model was applied for the quasi-neutral plasma flow acceleration in an ambipolar field. The coaxial structure of the ion thruster model is discussed. Preliminary estimate of the performance of the propulsion system prototype was carried out for a pulsed operation mode. Additionally, the energy budget of a high-frequency ion thruster prototype was calculated.

1. Introduction

The design of the propulsion system of the university nanosatellite is based at the parameters of it's motion [1,2]. A typical CubeSat nanosatellite is placed in the low Earth orbit [3,4]. It has orbital velocity of about 8 km/s. Typically the nanosatellite has the weight that is greater than 1 kg and the kinetic energy that exceeds 10^8 J. The propulsion system ability to do work must exceed this value in order to correct the nanosatellite orbit. Usually the energy flow that nanosatellite receives from the Sun is less then 100 W, therefore the orbit correction mission will take more then 10 days. Since the acceptable amount of propellant at the satellite is limited by value of 1 kg, the consumption rate of this substance must be less then 10^{-6} kg/s. The result is approximately 10 mN of thrust under assumption that it is created by protons accelerated up to 10-100 km/s. The concentration of particles at the nozzle about 10^{19} m^{-3} ensure the sufficient mean free path and the absence of particle collisions.

The model of the motion of charged particles in the electric and magnetic fields allow to understand operational principles of an ion propulsion system [5]. It is the first step of design procedure of the thruster. The charged particles tracing allow to estimate the jet thrust at the nozzle outlet and to propose the most effective construction of electric propulsion system [6-8].

A charged particle experiences a force perpendicular to its velocity when moving through a magnetic field. As a result it rotates in the plane perpendicular to the magnetic field with a frequency

Content from this work may be used under the terms of the Creative Commons Attribution 3.0 licence. Any further distribution of this work must maintain attribution to the author(s) and the title of the work, journal citation and DOI.
Published under licence by IOP Publishing Ltd

IOP Conf. Series: Materials Science and Engineering 1290 (2023) 012004 doi:10.1088/1757-899X/1290/1/012004

of ω_c. The frequency depends on the magnitude of the external magnetic field and is named cyclotron frequency. In the steady magnetic field, the charged particle rotates with a constant velocity along a circular orbit with a radius R_c. The frequency and radius of rotation is given by equations:

$$\omega_c = \frac{ZeB}{2\pi m};\ R_c = \frac{v_\perp}{\omega_c} \tag{1}$$

Where B- the magnitude of the magnetic flux density, Z- the particle charge number, $e = 1.602176634 \times 10^{-19}$ C - the elementary charge, $v_\perp$ -the velocity component orthogonal to the magnetic field, m- the particle mass. The disposition of the electric and magnetic fields determine the mechanisms of motion and the efficiency of particle acceleration.

It should be noted that properties of the magnetic field can be used to hold charged particle flow in order to design efficient accelerator. In the same time particle collisions is able to push out protons from a given volume and to prevent the circular motion under the influence of steady or divergent magnetic field. Therefore, the radius of rotation must be less than the particle's mean free path τ in order to perform a lot of rotations and to achieve effective acceleration. In other words the following condition must be satisfied:

$$\omega_c \tau = \frac{eB}{m v_c} >> 1 \tag{2}$$

It is necessary to apply an external electric field in order to accelerate the particle. The field can be constant E=E_0 or time-varying $E = E_0 \cos(\omega t)$.

2. Models of charged particle acceleration

In a fundamental sense, all existing models of particle acceleration are based on the same principle - the electric field accelerates the flux of particles, while the magnetic field determines the motion trajectory [9-10]. The electric field can be parallel to magnetic field or perpendicular. The mutual orientation of the electric and magnetic field determines two possible types of thrusters.

2.1. Acceleration of charged particles in conditions of parallel magnetic and electric fields.

The first model to be discussed is the acceleration of charged particles in parallel magnetic and electric fields. This type of acceleration is interesting for a number of reasons. The simplest example of such device is the linear accelerator [11]. A linear RF accelerator is a sequence of elements (grids or electrodes), each of which generates an oscillating electric field that match rotational frequency. The time interval during which the particles pass through the accelerating region between two elements is equal to the half-cycle of the electric field generator $T/2$. The phase of oscillations must be adjusted for each of the regions. The the distance between the accelerating regions (gaps between the electrodes) must be adjusted too in order to achieve maximum efficiency. The linear dimensions such accelerator type is tend to be relatively long if the high speed of charged particles is required. The simulation of particle motion in parallel fields is based on the following assumptions. The electric field is defined by the potential U applied between the electrodes. It oscillates with frequency ω: $U = U_0 \sin(\omega t + \varphi_0)$. The axial acceleration dv_x/dt of charged particles can be estimated by the equation:

$$\frac{dv_x}{dt} = \frac{eU_0}{md} \sin(\omega t + \varphi_0) \tag{3}$$

Each accelerating interval traveling distance must be equal to the distance between the electrodes required in order for the charged particle to pass single accelerating gap during the half of cycle $T/2$ (before the field direction reverses):

$$d = \frac{v_0 T}{4} + \sqrt{\left(\frac{v_0 T}{4}\right)^2 + \frac{eU_0}{m}\frac{T^2}{4\pi}} \tag{4}$$

The charged particles trace for the simplified model of a linear accelerator with a constant magnetic field of 0.1 T and the alternating electric field determined by a frequency of 300 kHz and a potential of 10 V is reflected at the figure 1.

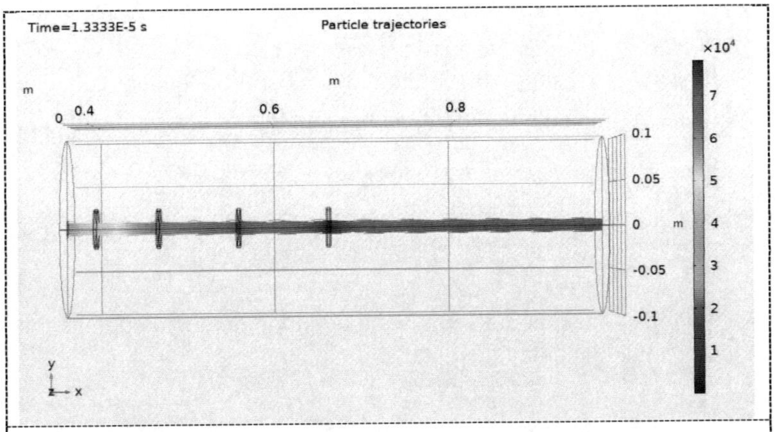

Figure 1. The charged particles trace for the model of linear accelerator

Given parameters determine the length of the accelerator which exceeds 60 cm. The increase in the accelerating potential and the frequency of the electric field oscillations allow to diminish the distance between the electrodes. It should be noted, that the acceleration efficiency of such coaxial structures can be improved by the focusing of the charged particle beam. The magnetic lens can be used to eliminate of beam divergence. The multipolar magnetic lens allow to keep charged particles away from the electrodes wall. The multipolar magnet was simulated with the help of COMSOL tools. The magnetic field configuration of multipolar magnet and particles traces are presented at the figure 2-3.

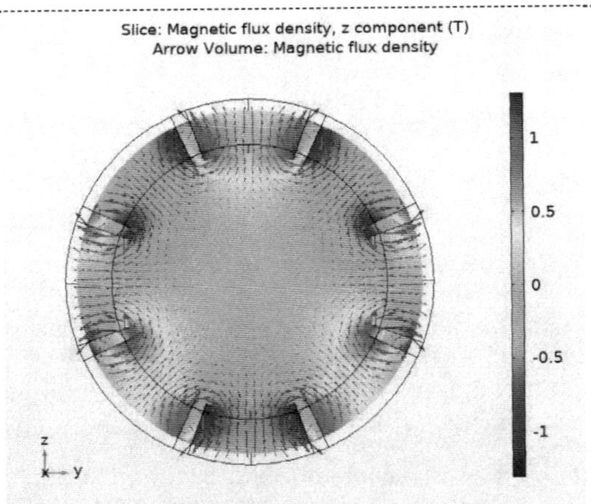

Figure 2. Multipolar magnetic field configuration of multipolar magnet

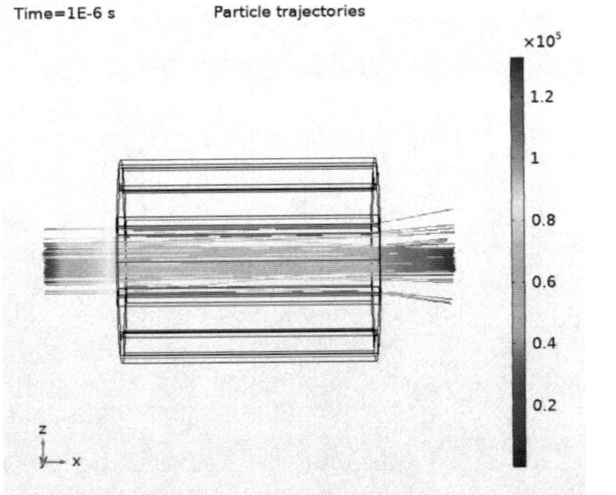

Figure 3. The charged particles tracing with magnetic confinement

IOP Conf. Series: Materials Science and Engineering 1290 (2023) 012004 doi:10.1088/1757-899X/1290/1/012004

2.2. Acceleration of charged particles in conditions of crossed magnetic and electric fields

The second model under discussion is the accelerator of particles by the crossed electric and magnetic field. In this case, the acceleration occurs due to the effect of cyclotron resonance. The frequency of electric field oscillations must be equal to the charged particle cyclotron frequency $\omega_c = \omega$. The beam passes through the accelerating field a lot of times and gains energy by small portions. It is the principle of classical cyclotron operation.

The COMSOL multiphysics tools were used to simulate the cyclotron acceleration of charged particles. The cylindrical crossed-field region with a constant magnetic and oscillating electric fields was used for the simulation. The charged particle trace is reflected at the figure 4. The simulation interval was equal 10 μs, magnetic field - 0.1 T and the oscillation frequency of the electric field - 9.36 MHz. The particle motion in crossed-field can be described by equations:

$$\begin{aligned} m\frac{dv_x}{dt} &= e(v \times B)_x = ev_y B \\ m\frac{dv_y}{dt} &= e[E + (v \times B)]_y = eE - ev_x B \\ m\frac{dv_z}{dt} &= 0 \\ F_e &= eZ\,\mathrm{Re}\left(Ee^{j(\omega t + \varphi_0)}\right) \end{aligned} \tag{5}$$

Where F_e - the force on a charged particle between acceleration plates.

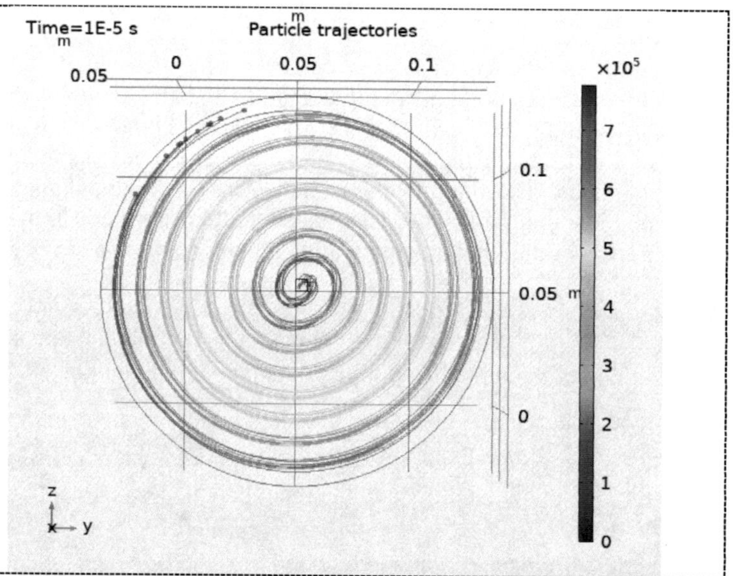

Figure 4. The charged particles tracing in the cyclotron acceleration model

The effective technique of beam extraction from the region of acceleration is required when the design of a cyclotron for accelerating charged particles is carried out. Beam can be deflected with the help of additional source of static magnetic field. One of the possible technique can be illustrated by the example of the charged particles acceleration in the conditions of divergent magnetic field [12-14]. Such configuration of fields allows to create the thruster geometry of the coaxial form.

The simplified geometry for charged particles tracing in the divergent magnetic field include the source of charged particles, permanent magnets setup and the oscillating electric field. The motion of charged particles in the conditions of crossed magnetic and electric fields can be described by equations:

$$m\frac{dv}{dt} = qE + q[v \times B] \tag{6}$$

The results of simulation of the thruster which utilize the acceleration of charged particles in a divergent magnetic field is shown at the figure 5-6. The simulation shows that axial electric field accelerates particles mainly during their rotation in the conditions of uniform magnetic field in the middle of the magnetic system. After the sufficient increase of the rotation radius the particle is reflected. In other words, the divergent magnetic field transforms vector of radial velocity into a longitudinal one.

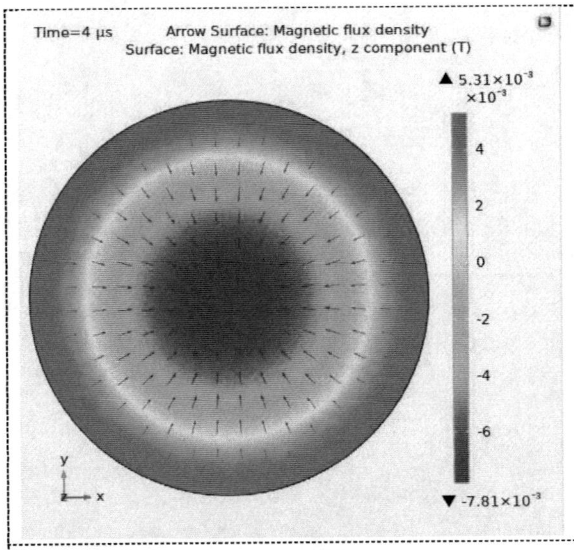

Figure 5. Divergent magnetic field configuration

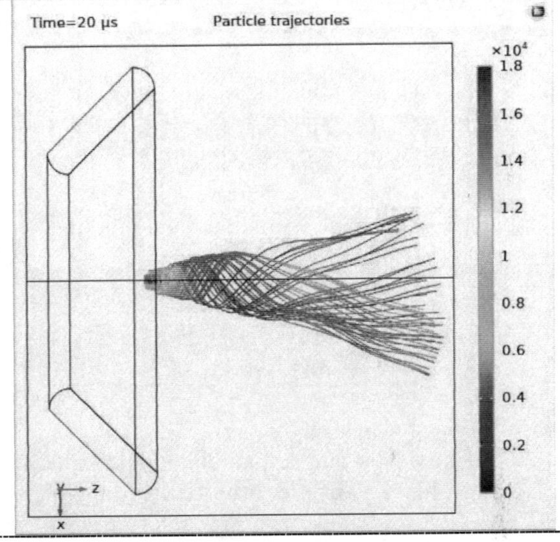

Figure 6. Charged particles tracing in the divergent magnetic field

3. Prototyping of the ion thruster

The desire to minimize the weight and size of the ion thruster explains the enthusiasm concerning electrodeless engine designs. Also this way allow to prevent rapid degradation of thruster elements. Perhaps the exclusion of the electrodes is not the best solution. It is more reliable to form the structure of electrodes which allow to minimize the possibility of direct impact of the charged particle. The principle of linear acceleration can satisfy the conditions of minimal interaction with charged particles. The ring or cylindrical geometry of electrodes allow to realize this principle. The problem of neutralization is solved there by placing the zero potential electrode at the end of accelerating structure.

One can propose the structure of two spirals with a variable pitch as a more attractive and technically advanced form of electrodes implementation. The pitch of the spiral is equivalent to gap between the electrodes of linear accelerator. The trajectory to be followed by charged particles under the influence of a magnetic and electric field is expected to follow the shape of the spiral, while maintaining quasi-neutrality of due to electrodynamics forces.

IOP Conf. Series: Materials Science and Engineering 1290 (2023) 012004 doi:10.1088/1757-899X/1290/1/012004

The prototype of the linear RF accelerator for the nanosatellite thruster was simulated under the following conditions: the total size of the accelerating region is about 10 cm, the required velocity of the output beam is 10–100 km/s, the potential between spirals is V=10 V, and the electric field oscillation frequency of 1.52 MHz The charged particles tracing for the thruster with electrodes in the form of two spirals is shown at the figure 7-8. The charged particles interaction is reduced thereof propellant utilization efficiency must be improved. Thus, during the process of thruster design, the structure of the thruster can be optimized to ensure unhindered movement of charged particles through the operational geometry.

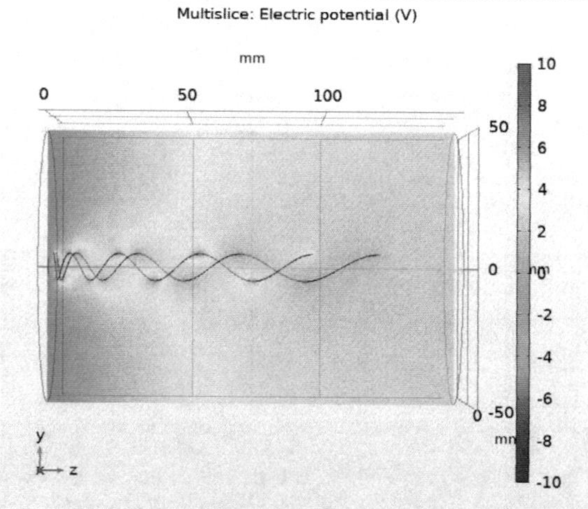

Figure 7. The electric field potential configuration

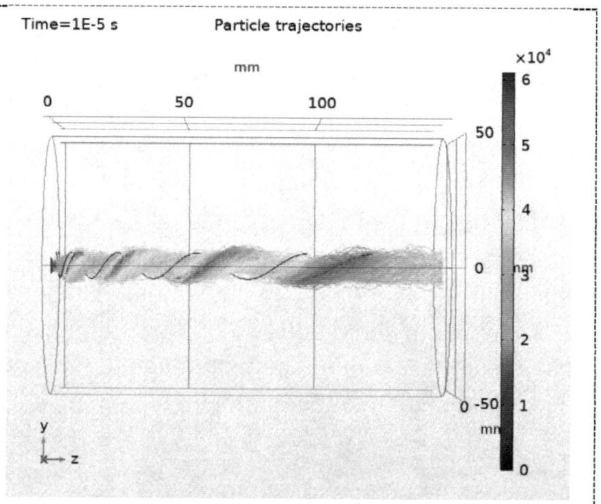

Figure 8. Charged particles tracing in the geometry of a linear RF accelerator

4. Ion thruster design

The performed simulations of charged particle accelerators allow to propose two possible schematic diagrams of an ion thruster prototype for a university nanosatellite.

The first one is based on the effect of cyclotron resonance in crossed electric and magnetic fields. The structure of the thruster is presented at the figure 9.

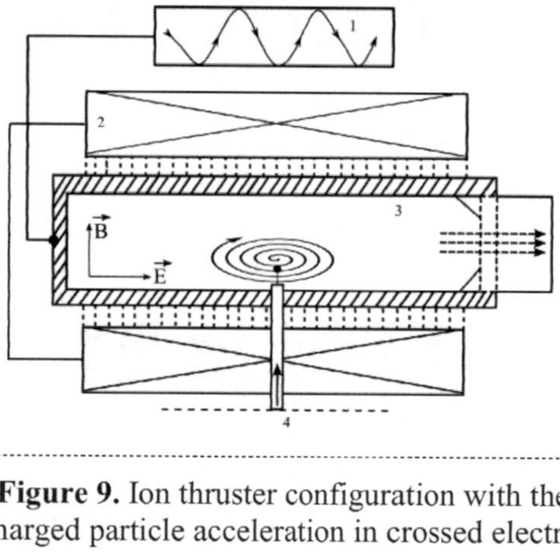

Figure 9. Ion thruster configuration with the charged particle acceleration in crossed electric and magnetic fields

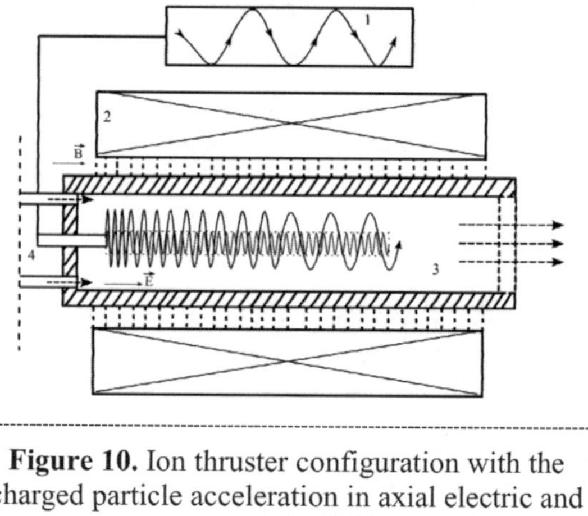

Figure 10. Ion thruster configuration with the charged particle acceleration in axial electric and magnetic field

1-RF source circuit, 2- permanent magnets, 3- acceleration region, 4- propellant injection

The second utilize acceleration in parallel electric and magnetic field. The ion thruster structure has a coaxial form. It is based on the principle of linear acceleration. The electrodes are implemented in the form of two spirals with a variable pitch equivalent to the gap between the electrodes of linear accelerator. The schematic diagram of the coaxial ion thruster is shown at the figure 10.

5. Conclusion

The paper presents the results of development of a neutralizer-free high-frequency ion thruster for university nanosatellite propulsion systems which allow it to perform orbit correction tasks. The simulations of a number of charged particle accelerators were performed. The analysis of plasma flow and the charged particles acceleration was carried out with the COMSOL Multiphysics package. The results allow to propose two possible schematic diagrams for prototype of an ion thruster of a university nanosatellite. The first one is based on the effect of cyclotron resonance in the crossed electric and magnetic fields. The second utilize acceleration in parallel electric and magnetic field. It is based on the principle of linear acceleration. The electrodes are implemented in the form of two spirals with a variable pitch equivalent to the gap between the electrodes of linear accelerator.

References

[1] O'Reilly, D.; Herdrich, G.; Kavanagh, D.F. Electric Propulsion Methods for Small Satellites: A Review. Aerospace 2021, 8, 22. https://doi.org/10.3390/aerospace8010022
[2] I. Levchenko, S. Xu, S. Mazouffre, D. Lev, D. Pedrini, and et. , "Perspectives, frontiers, and new horizons for plasma-based space electric propulsion", Physics of Plasmas 27, 020601 (2020) https://doi.org/10.1063/1.5109141
[3] K. Lemmer, "Propulsion for CubeSats", Astronautica , vol.134, pp.231-243, 2017.
[4] Tripathy, P., "Overview On Electric Propulsion Systems", International Journal of Scientific and Research Publications, vol. 10, № 12, pp. 422-439, 2020.
[5] K. Holste, P. Dietz, S. Scharmann, K. Keil, T. Henning, D. Zschätzsch, and et., "Ion thrusters for electric propulsion: Scientific issues developing a niche technology into a game changer", Review of Scientific Instruments 91, 061101 (2020) https://doi.org/10.1063/5.0010134
[6] Yang J., Jia S., Zhang Z. et al. 2020 Performance of a 4-cm iodine-fueled radio frequency ionthruster Plasma Science and Technology (vol. 22, No 9) pp 1–15
[7] Kramer, A.; Bangert, P.; Schilling, K. UWE-4: First Electric Propulsion on a 1U CubeSat—In-Orbit Experiments and Characterization. Aerospace 2020, 7, 98. https://doi.org/10.3390/aerospace7070098
[8] Nakagawa, Y., Koizumi, H., Naito, Y., and Komurasaki, K., "Water and xenon ECR ion thruster—comparison in global model and experiment", Plasma Sources Science Technology, vol. 29, no. 10, 2020. doi:10.1088/1361-6595/aba2ac.
[9] Goebel D M and Katz I (2008) Fundamentals of Electric Propulsion: Ion and Hall Thrusters (New York: Wiley)
[10] Reece Roth, J. (1995). Industrial Plasma Engineering: Volume 1: Principles (1st ed.). CRC Press. https://doi.org/10.1201/9780367802615
[11] Vretenar, M., "Linear Accelerators", arXiv e-prints, 2016. doi:10.48550/arXiv.1601.05210.
[12] Lafleur T, Cannat F, Jarrige J, Elias PQ, Packan D (2015) Electron dynamics and ion acceleration in expanding-plasma thrusters. Plasma Sources SciTechnol 24:065013. https://doi.org/10.1088/0963-0252/24/6/065013
[13] Porto, J. C., and Elias, P.-Q., "Full-PIC Simulation of an ECR Plasma Thruster with Magnetic Nozzle," 36th Int. Electr. Propuls. Conf., Vienna,Austria, 2019, pp. IEPC–2019–232.
[14] F. Cannat, T. Lafleur, J. Jarrige, P. Chabert, P.-Q. Elias, and D. Packan , "Optimization of a coaxial electron cyclotron resonance plasma thruster with an analytical model", Physics of Plasmas 22, 053503 (2015) https://doi.org/10.1063/1.4920966

IOP Conf. Series: Materials Science and Engineering 1290 (2023) 012005 doi:10.1088/1757-899X/1290/1/012005

Retrofit of a standard split-type air conditioner into an air-to-water heat pump

Boţu Mădălina[1], Atanasiu Marius[2], Ursescu Gabriel[2], Rakosi Edward[2]

[1] National Research & Development Institute for Gas Turbine COMOTI, Bucureşti
[2] "Gheorghe Asachi" Technical University of Iaşi

madalina.botu@student.tuiasi.ro

Abstract. In the context of the rising prices of energy and especially of natural gas, the interest in heat pumps has grown steadily over the last period, as they are a very convenient alternative to low and medium power thermal plants. Although the price of heat pumps has dropped a lot recently, these systems are still very expensive compared to the wall-mounted natural gas fuelled power plants found in the vast majority of residential buildings in Romania. In this context, the authors proposed the conversion of an inverter type air conditioner into an air-water heat pump with minimal costs, maintaining and even improving the global performance of the system. The newly made heat pump was tested at variable regimes in the Cogeneration Trigeneration Laboratory of the Faculty of Mechanical Engineering of Iasi. The external unit was inserted into a climatic chamber capable of simulating variable climatic conditions for any region in Europe. Later, the same heat pump was tested in real operating conditions, being inserted in the heating circuit of a medium-sized house in the Iasi area. Its operation was then monitored during an entire winter season, being instrumented to measure the energetic balance and determine the coefficient of performance (COP) continuously throughout the season, in accordance with the weather conditions and also to measure the seasonal coefficient of performance (SCOP).

1. Introduction

Heat pumps are an excellent alternative to provide comfort in small households compare to fossil fuel boilers. They require less to no infrastructure investment since electrical grid is fairly good almost every ware. Beside that heat pumps are very easy to control and automate via internet. This day most of the heat pumps available on market can be controlled remotely. The maintenance of a heat pump system is much cheaper and easier to do compared to any another heating system. Taking into account these advantages, heat pumps have seen an accelerated development in the last period. Another factor that accelerates the heat pumps implementation, expressly in European Union, is National Energy and Climate Plan which details a strategy to reach 32% renewable energy by 2030 [01].

However, heat pumps are still a rarity, especially on the Romanian market. The main disadvantages are related to the modest performance of these systems when implemented in radiator heating systems, as well as their price, which is still quite high. Some studies conducted in European countries show a decrease in COP values to 1.4 [02] or even lower for air-water heat pumps implemented in old systems with thermal agent distribution through radiators.

A relatively cheap alternative would be to use traditional air conditioning systems that can be used for heating in old houses that already have a classic heating system. The use of air conditioning, especially when the outside temperatures are positive, can greatly reduce heating costs. For example, in Romania,

Content from this work may be used under the terms of the Creative Commons Attribution 3.0 licence. Any further distribution of this work must maintain attribution to the author(s) and the title of the work, journal citation and DOI.
Published under licence by IOP Publishing Ltd

any heat pump with a COP greater than 2 produces thermal energy at a lower cost than the best thermal power plant, assuming the best electricity and natural gas prices [03], [04] available on the Romanian market.

Heat pumps are still attractive even for newly built houses where the methane gas distribution infrastructure is non-existent or poorly developed. In this case, the implementation of a heat pump is not as expensive if the costs for building the methane gas network are also taken into consideration.

2. AC retrofitting description and operation

To reduce the costs as much as possible with such an investment, the authors proposed the conversion of a conventional air conditioner into an air-to-water heat pump. The main motivation for this conversion was the fact that the cheapest heat pump with a power of 3.5 kW is over three times more expensive than an air conditioner with equivalent power. This price discrepancy remains for all small power heat pumps (under 10 kW) when compared to conventional air conditioners.

Thus, to transform the air conditioner into a heat pump, the authors used the heat exchanger from the indoor unit, which was placed inside a metal cylinder. The refrigerant circulates inside the coils of the heat exchanger as in a conventional air conditioner, while outside the heat exchanger is washed with water instead of air, as shown in figure 1. Considering the very large surface area of the heat exchanger and taking into account the significant increase in the heat transfer coefficient by convection to water compared to air, the new system thus created has even better performance than a conventional air conditioner.

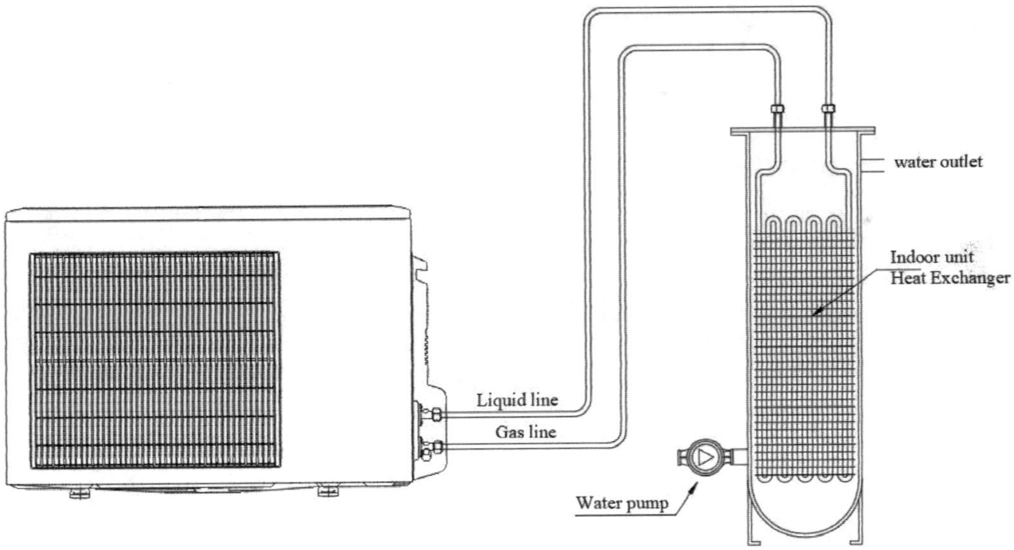

Figure 1. The diagram of the air conditioning system converted into an air-to-water heat pump.

2.1. *Integration of the installation into the heating scheme of a house*

For performance estimation in real operating conditions, the system was installed on the heating circuit of a single-family home located in the outskirts of the city of Iasi. The house is a ground plus first floor structure, with a built area of 72 square meters and a living area of 115 square meters. The construction structure is made of reinforced concrete, and the walls are made of BCA with a thickness of 400 mm and are insulated on the exterior with 100 mm thick expanded polystyrene.

IOP Conf. Series: Materials Science and Engineering 1290 (2023) 012005 doi:10.1088/1757-899X/1290/1/012005

The heating is done through the floor, both on the ground floor and in the attic. The heat transfer agent is water, which circulates through a total of 12 floor coils, 7 on the ground floor and 5 in the attic. From an automation standpoint, the system is divided into 5 zones, of which 4 are assisted by wall thermostats that operate control valves for the corresponding heating coils' flow rate. The fifth zone is only controlled by the heating circuit's automation.

In the initial design, the heat source was a 12 kW wall-mounted electric heating system. Besides heating, this system also provides domestic hot water. This is prepared in a 200-litre boiler equipped with 2 coils through which the hot water produced in the electric heating system circulates. The agent's circulation is ensured by the internal pump, and the distribution of heat to the boiler or heating circuit is controlled by a 3-way valve. The automation prioritizes domestic hot water, so during its preparation, all the thermal energy produced is sent to the water boiler.

In order to integrate the heat pump into the heating system, a series of modifications were made to the circuits so that the heat transfer agent could also be heated by the heat exchanger of the heat pump condenser. The method of integrating this heat exchanger into the system is presented in the figure 2.

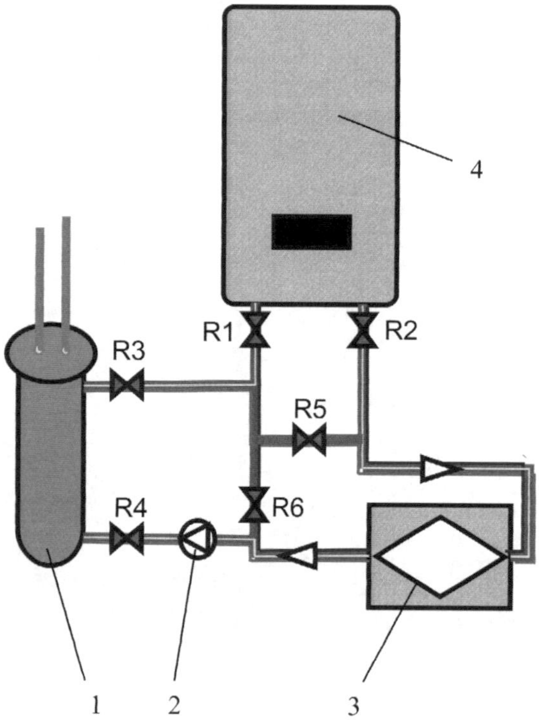

Figure 2. Integration of the heat pump into the heating scheme of a house

1 – Indoor unit heat exchanger
2 – Water pump
3 – Internal heating circuit (floor water coils)
4 – Electric heating system

The implemented scheme allows for operation in several modes, depending on the position of the valves (R1...R6). Thus:

- the heat pump can operate independently as the sole source of heat for the house. In this case, R1, R2, and R6 will be closed, while R3, R4, and R5 will be open. The circulation of the heat transfer agent will be ensured by the additional pump (position 2 in figure 2).
- the electric heating system can operate independently (R1, R2, and R6 open, R3, R4, and R5 closed);
- the heat pump and the electric system can operate in series (R1, R2, R3, and R4 open, R5 and R6 closed). Circulation will be provided by either the internal pump or both pumps.

The last operating mode is useful on days with very low outside temperatures to compensate for additional heat losses that the heat pump, due to its relatively low power, cannot cover. In fact, during

IOP Conf. Series: Materials Science and Engineering 1290 (2023) 012005 doi:10.1088/1757-899X/1290/1/012005

the season in which the measurements were taken, it was necessary to operate in this mode on several occasions to maintain the indoor temperature within the required limits.
It should be noted that the domestic hot water is still prepared only by the electric heating system, which operates whenever necessary.

3. Experimental procedure

3.1. Equipment description
The measurements carried out during the experiments aimed to establish the energy balance of the heat pump, in the 2 configurations studied: during testing in the climatic chamber and during testing in real operating conditions. To measure the energy and power consumed by the heat pump, a Voltcraft Energy Logger 4000 type electrical energy meter was used, which allows measurement of the current consumed parameters (voltage, current, frequency, cos ϕ). It was mounted in the socket that supplies the heat pump. In addition to these parameters, the device measures the instantaneous power consumed and also integrates the energy consumed up to the reading time. To establish the energy balance, it was also necessary to measure the thermal energy produced by the heat pump. For this purpose, a Sharky 775 type thermal energy meter was used, which measures the following parameters: water temperature at the inlet of the circuit, water temperature at the outlet, and water flow rate. Based on these, the device calculates and displays the thermal power and also integrates the thermal energy produced up to the reading time. The thermal energy meter was installed to measure the water flow rate at the outlet of the heat exchanger of the pump condenser. The temperature sensors are mounted, one at the inlet and the other at the outlet of the same heat exchanger. The installation mode of the energy meters is shown in the figure 3.

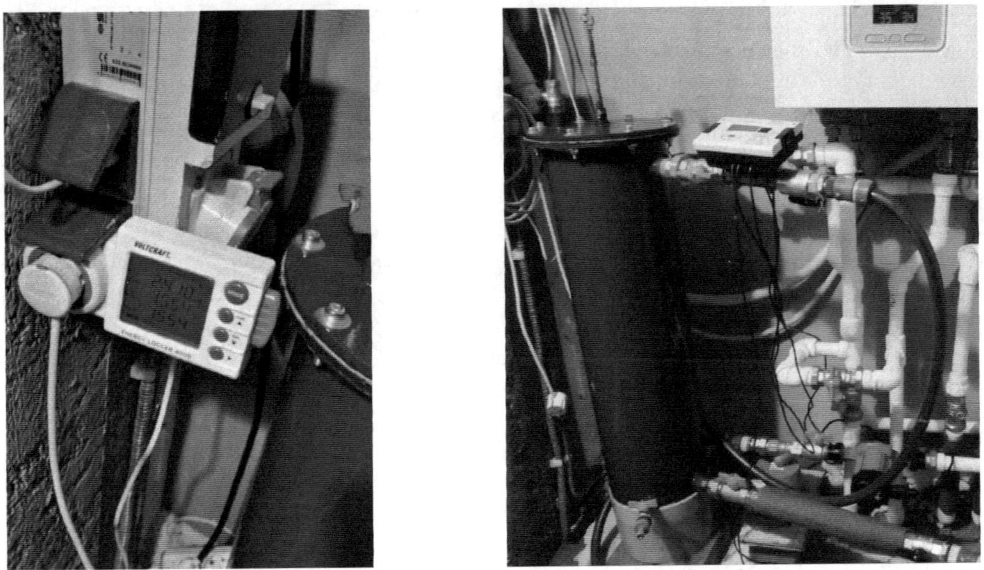

a b

Figure 3. Energy meters for electrical energy consumed (a) and thermal energy produced (b)

3.2. Measurements in the climatic chamber
The system was initially installed in the Faculty of Mechanics laboratory in Iasi. Figure 4 shows pictures taken during the laboratory tests. The outdoor unit was installed in a climatic chamber where temperature and humidity could be controlled. Following tests under controlled conditions, the following COP values were obtained:

- At a temperature of -5°C set in the climatic chamber, the obtained COP was 2.01

- At a temperature of 0°C set in the climatic chamber, the obtained COP was 2.32
- At a temperature of 5°C set in the climatic chamber, the obtained COP was 2.96

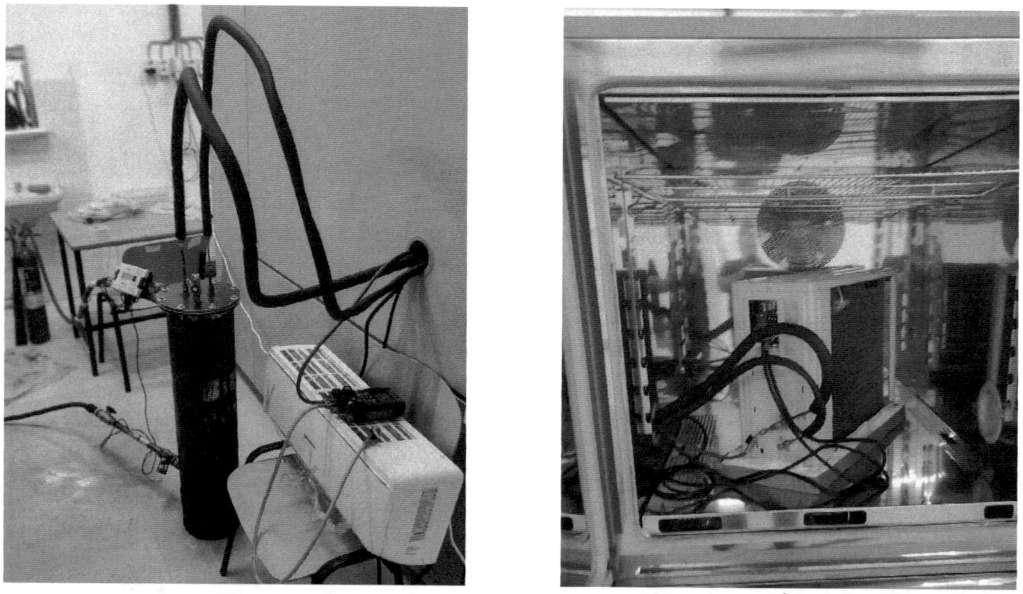

a b

Figure 4 The indoor unit of the heat pump (a) and the outdoor unit of the heat pump (b).

It should be added that during the laboratory testing, the humidity in the climatic chamber was not controlled, and as a result, the actual relative humidity in the chamber was very low. The relative humidity measured at -5°C was below 40%, which is why the formation of ice on the external unit's heat exchanger was practically non-existent.

3.3. In-situ measurements

The in-situ measurements were carried out from December 19th, 2021 to April 23rd, 2022, covering practically an entire winter season. The installation site is located in the village of Ciurbesti, Miroslava commune, Iasi County, at 47.09° north latitude and 27.54° east longitude. The area is characterized by a moderate continental climate. The season in which the measurements were made can be considered representative of the climate of the respective area. During the measurements, the thermostats in the house were set to a temperature of 22°C. The heat pump was set so that the return water temperature was maintained at a minimum of 27°C. Depending on the outside temperatures, the supply water temperature varied between 32-37°C during the measured season. The measurements were taken at a frequency of at least once a day, except for a short period of 3 days when the measurements could not be taken. The following parameters were recorded at each measurement:

- The time of the measurement [date and time]
- The index of the thermal meter [MJ]
- The index of the electric meter [kWh]

In addition, the temperature and humidity of the air were recorded at the time of the measurement. For recording these, the data from the website https://www.wunderground.com/ [5] was used, which presents weather data using the records of different weather stations in the territory. In our case, the data provided by the station in the locality of Horpaz, located at a distance of 800 m from the installation, were used. Based on the measured data, the following parameters were then calculated:

- The thermal energy produced by the heat pump between two consecutive readings - E_{util} [MJ]: it is equal to the difference between two consecutive readings.

- The thermal energy produced - E_{util} [kWh]:

$$E_{util}\ [kWh] = E_{util}\ [MJ] \cdot 0{,}277778$$

- The electric energy consumed since the last reading - E_{cons} [kWh]: it is equal to the difference between two consecutive readings.
- Coefficient of performance of the heat pump - COP.

$$COP = \frac{Eutil\ [kWh]}{Econs\ [kWh]}$$

- Time interval between 2 readings - Δt [hours]
- Average thermal power produced by the heat pump in the last measured interval - $P_{t\ medie}$

$$P_{t\ medie}\ [kW] = \frac{Eutil\ [kWh]}{\Delta t\ [ore]}$$

- Average electrical power consumed in the last measurement interval - $P_{e\ medie}$ [kWh].

$$P_{e\ medie}\ [kW] = \frac{Econs\ [kWh]}{\Delta t\ [ore]}$$

These data have been collected in the table presented below. In order to save space, only the first and last measurements are presented in table 1, although a total of 172 measurements were made throughout the season.

Table 1. Experimental results in situ.

Data	Ora	T_{ext}	umid (%)	T_{med}	Index Q	E_{util} [MJ]	E_{util} [kWh]	Index E	E_{cons} [kWh]	COP	Dt [ore]	$P_{t\ medie}$ [kW]	$P_{e\ medie}$ [kW]
18.12.2021	19:40	1,2	85	-	897	0	0,00	0,00	0	-	-	-	-
18.12.2021	22:01	1,6	84	1,4	921	24	6,67	2,41	2,41	2,766	2,35	2,84	1,03
18.12.2021	23:46	1,7	84	1,65	940	19	5,28	4,26	1,85	2,850	1,75	3,02	1,06
19.12.2021	7:46	2	85	1,85	1032	92	25,56	13,05	8,79	2,908	8,00	3,19	1,10
19.12.2021	9:31	1,7	89	1,85	1046	14	3,89	14,34	1,29	3,015	1,75	2,22	0,74
19.12.2021	11:14	3,7	84	2,7	1060	14	3,89	15,54	1,20	3,241	1,72	2,27	0,70
19.12.2021	13:58	6,5	76	5,1	1091	31	8,61	17,98	2,44	3,529	2,73	3,15	0,89
19.12.2021	22:18	5,1	75	5,8	1168	77	21,39	24,14	6,16	3,472	8,33	2,57	0,74
20.12.2021	9:30	0,6	84	2,85	1265	97	26,94	32,57	8,43	3,196	11,20	2,41	0,75
20.12.2021	19:53	1,8	79	1,2	1371	106	29,44	41,88	9,31	3,163	10,38	2,84	0,90
—	—	—	—	—	—	—	—	—	—	—	—	—	—
5.4.2022	23:53	9,3	58	6,1	23121	275	76,39	2146,00	24,00	3,183	36,95	2,07	0,65
7.4.2022	9:02	11,3	59	10,3	23256	135	37,50	2157,00	11,00	3,409	33,15	1,13	0,33
10.4.2022	11:27	4,6	91	12	23511	255	70,83	2176,00	19,00	3,728	74,42	0,95	0,26
11.4.2022	8:47	3,9	80	4,25	23625	114	31,67	2186,00	10,00	3,167	21,33	1,48	0,47
17.4.2022	22:21	4,3	71	12	24575	950	263,89	2264,00	78,00	3,383	157,57	1,67	0,50
20.4.2022	21:28	7,6	76	9	25034	459	127,50	2304,00	40,00	3,188	71,12	1,79	0,56
23.4.2022	20:19	11,8	89	9,7	25367	333	92,50	2330,00	26,00	3,558	70,85	1,31	0,37

Based on the tabulated data, the variation curves of the following parameters were generated: exterior temperature, exterior relative humidity, electric power consumption, thermal power produced, and coefficient of performance. These curves were represented together in the graph in the figure 5.

4. Result interpretation

After analyzing the experimental results presented in the graph in figure 5, we can state the following:

- The average outside temperatures during the measured season varied between -7,7° C and 16,35° C.
- During the studied season, the coefficient of performance of the system varied between 2.05 and 4.63.

IOP Conf. Series: Materials Science and Engineering 1290 (2023) 012005 doi:10.1088/1757-899X/1290/1/012005

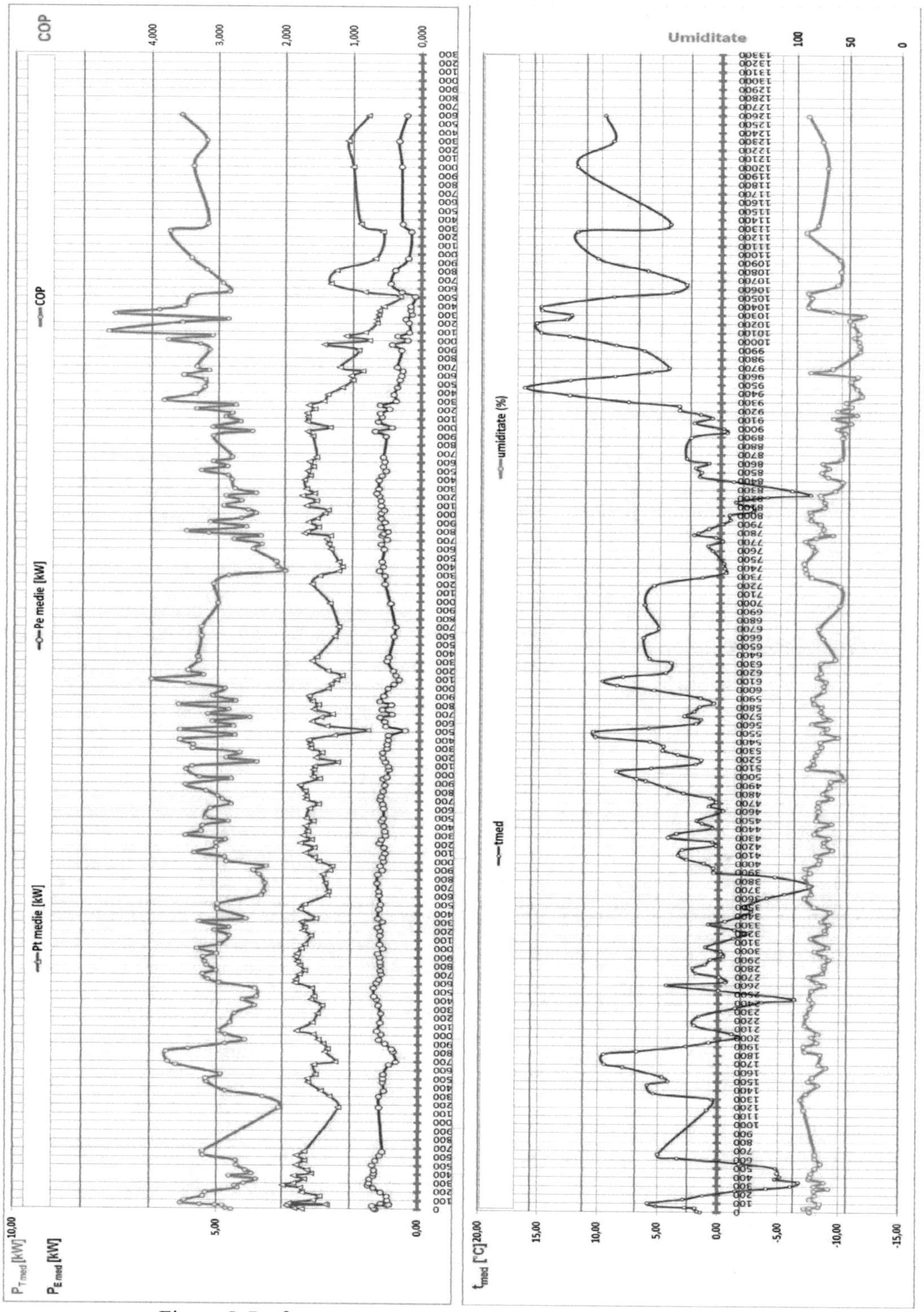

Figure 5. Performance parameters variation throughout the season

- The variation of COP was due to multiple factors. Analyzing the curves in figure 5, one can observe that the variation of COP is closely related to the variation of the external temperature, the two curves having a very similar shape. However, there are small areas in which the variation of COP does not strictly follow the variation of the external temperature. This was usually due to the clogging of the external heat exchanger caused by frost or snow cover. Such a case can be observed during the period of December 26-31, 2021 (days 7 to 12 of the observation period), when the system operated unsupervised and at the end of which the outdoor unit was found covered with a thick layer of snow.
- The efficiency of the heat pump is influenced by weather conditions. In this regard, a series of "anomalies" in operation can be observed on the diagrams. For example, we observe from the diagram that the pump efficiency is very low around the 73rd day of measurement (March 2, 2022), although the outdoor temperature is around 0°C, a temperature at which, in other situations, the pump efficiency was much higher. If we look at the curve of humidity variation during this period, we observe that it had very high values. Moreover, during the measurements, a massive ice obstruction of the outdoor unit heat exchanger was recorded in this period. Therefore, the decrease in efficiency is not due to the outdoor temperature, but to the fact that the outdoor unit operated obstructed due to the frost produced by humidity, thus being unable to extract enough heat from the outside environment. This phenomenon is confirmed by the curve of the thermal power produced, which has a low value, although the electric power consumed is relatively high. This operating situation was encountered several times during the season and is one of the problems that should be addressed to optimize the operation of the installation.
- The average electrical power consumed by the installation varied between 0,16 kW and 1,22kW, while the average thermal power produced varied between 0,53 kW and 3,30 kW. It is observed that, as expected, both the electrical power consumed and the electrical power produced increase when the outdoor temperature drops and decrease on warm days.
- The electric energy consumed throughout the season is 2330 kWh. Meanwhile, the heat pump produced a quantity of 7148 kWh of thermal energy, which leads to a seasonal coefficient of performance SCOP = 3,068.
- It is observed that the efficiency of the heat pump in situ was generally higher than the one measured in the laboratory. This fact can be attributed to the frosting of the external heat exchanger, which, in the early phase, can lead to an increase in heat exchange and COP [6]. As mentioned earlier, when the ice layer becomes too thick, it affects the air circulation through the exchanger, which leads to a rapid decrease of the heat extracted from the outside environment. In most of the cases, the automation of the installation correctly detected the thickening of the ice layer and promptly intervened for de-icing.
- The average thermal power during the hole season was 2,23 kW, while the average electrical power consumed during the same period has 0,764 kW.

5. Conclusion

Heating with heat pumps is a technology with a huge development potential around the world and especially in European Union since a lot of laws and regulation implies more actions to reduce CO2 emissions. Regarding to the dynamic of fossil fuel and electricity price, heat pumps are becoming more interesting since the increasing of electricity price is slower and les variable compared to the price variation of natural gas. An important role of electricity price is played by the increasing number of solar panel installation those who contribute to a more stable electricity price and to an important lowering of CO_2 emissions.

Since the prices of heat pumps are still too high, we proved that the transformation of a conventional air conditioning system into an air to water heat pump is possible with minimum investments and with notable benefits if we take into account the increase of refrigerant subcooling and a slight increase of the SCOP.

Our work demonstrates that a relatively efficient air-to-water heat pump can be made with minimal costs and low technology by modifying an existing air conditioner with inverter technology. Considering the SCOP obtained in this winter season, representative for the Iasi area, it results that the heating with this heat pump was cheaper than heating with natural gas if we take into account current energy prices.

References

[1] Chesser M, Lyons P, O'Reilly P, Carroll P, 2021, Air source heat pump in-situ performance *Energy & Buildings* **251** 111365 https://doi.org/10.1016/j.enbuild.2021.111365

[2] Carroll P., Chesser M., Lyons P., 2020, Air Source Heat Pumps field studies: A systematic literature review *Renewable and Sustainable Energy Reviews* **134** 110275 https://doi.org/10.1016/j.rser.2020.110275

[3] https://www.anre.ro/ro/info-consumatori/comparator-oferte-tip-de-furnizare-a-gn

[4] https://www.anre.ro/ro/info-consumatori/comparator-de-tarife

[5] www.wunderground.com

[6] Guo X, Chen Y, Wang W, Chen C, 2008, Experimental study on frost growth and dynamic performance of air source heat pump system, *Applied Thermal Engineering* **28** 2267–2278

IOP Conf. Series: Materials Science and Engineering 1290 (2023) 012006 doi:10.1088/1757-899X/1290/1/012006

Presentation of some constructive solutions for a rotating machine with profiled rotors

Gabriel Fischer-Szava[1], Alexandru Dobrovicescu[1], Georgiana Dăescu (Duiculete)[1], Nicolae Băran[1], Mihaela Constantin[1,2], Cătălina Dobre[1,2], Rana Adil Abdul-Nabe[1], Sevastiana Areta Ghioca[1]

[1]University Politehnica of Bucharest, Faculty of Mechanical Engineering and Mechatronics, Bucharest, Romania
[2]Academy of Romanian Scientists, Ilfov 3, 050044, Bucharest, Romania

i.mihaelaconstantin@gmail.com

Abstract. The purpose of the paper is to present an original solution for conveying some liquids with the help of a rotating volumetric pump. The pump can be equipped with 1 to 4 specially processed profiled rotors and can transport viscous fluids, pure fluids, or polyphase fluids. It was concluded that the performance of this machine, which can also be used as a fan or low-pressure compressor, is superior to other similar machines.

1. Introduction

Due to the limitations related to the current development requirements in the industry, for example maintaining a high volumetric efficiency in different operating conditions of the traditional piston pumps, a new rotary pump with profiled rotors is proposed in the paper.

In this study, a new pump with three or four profiled rotors as well as the operation scheme for increasing the performance were comprehensively introduced.

The main characteristics that must be considered in the design of a pump are related to the smallest possible size and high performance (e.g., high pressure, high rotation speed) [1][2].

Axial piston pumps are the most widely used hydraulic pumps. The specialized literature offers a complete description of their operating principle [3]. Also, their efficiency is studied in the paper [4]. Since the performance of these types of pumps, energy losses and their volumetric efficiency are very important in the construction of a pump, it have been intensively studied by researchers in the field [5–7]. However, standard piston pumps use a limited number of pistons for oil absorption and drainage [8]. Many researchers have tried in recent years to modify the structure and characteristics of pumps so that their performance is as good as possible [9][10].

In the present study, this paper proposes to design a new rotating volumetric pump with 3 or 4 profiled rotors of original shape that can be used to further increase the power and volumetric efficiency.

2. Materials and Methods

The performance increase can take place by optimizing the internal architecture of these pumps; there are different shapes both for the rotating pistons and for the shape of the rotors [11][12].

Content from this work may be used under the terms of the Creative Commons Attribution 3.0 licence. Any further distribution of this work must maintain attribution to the author(s) and the title of the work, journal citation and DOI.
Published under licence by IOP Publishing Ltd

IOP Conf. Series: Materials Science and Engineering 1290 (2023) 012006 doi:10.1088/1757-899X/1290/1/012006

To design the rotors, a calculation program was created and later, the processing of the rotors was carried out on a numerical processing center [13].

To determine the flow rate transported by the machine that would contain a single rotor, because the flow rate transported by the rotating machine is determined by the shape of the rotor, more precisely by the shape of the rotating piston, the machine in figure 1 is considered.

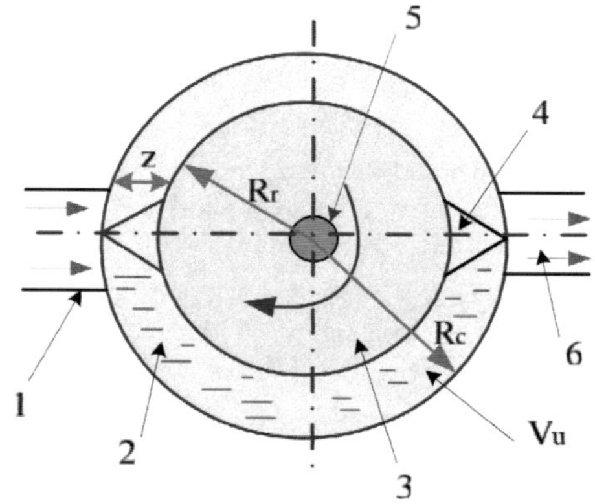

Figure 1. Rotating machine. 1- suction chamber, 2- case, 3- rotor, 4- rotating piston, 5 -shaft, 6- discharge chamber

It is denoted by V_u, the volume of fluid contained between the pistons and the case.

At a 360° rotation of the rotor 3, two volumes V_u will be transported from suction to discharge; in this case, the volumetric flow will be:

$$V_u = 2\left(\frac{\pi R_c^2}{2} - \frac{\pi R_r^2}{2}\right)\cdot l \;\left[m^3/rot\right] \tag{1}$$

In equation (1), the case radius (R_c) will be the sum of the rotor radius (R_r), rotor length l [m] and the piston height (z).

$$R_c = R_r + z \tag{2}$$

Substituting equation (2) in (1), it results:

$$V_u = \pi z l (z + 2R_r) \;\left[m^3/rot\right] \tag{3}$$

A single rotor of length l [m] and speed n_r [rot/min] will deliver a flow rate of:

$$\dot{V}_u = \pi z l (z + 2R_r)\frac{n_r}{60} \;\left[m^3/s\right] \tag{4}$$

The volumetric flow rate ($\dot{V}_m$) transported by the rotating pump with two rotors will be:

$$\dot{V}_m = 2\dot{V}_u = \pi z l (z + 2R_r)\frac{n_r}{30} \;\left[m^3/s\right] \tag{5}$$

The theoretical flow rate of the pump with triangular pistons will be reduced by V_P - the volume of equal prisms that give the volume of a piston of triangular shape, resulting:

$$\dot{V}_m = \pi z l (z + 2R_r)\frac{n_r}{30} - V_p \;\left[m^3/s\right] \tag{5'}$$

IOP Conf. Series: Materials Science and Engineering 1290 (2023) 012006 doi:10.1088/1757-899X/1290/1/012006

From equation (5), one can see that with the increase of l, z, R_r, n_r, the volumetric flow rate transported by the pump will increase.

To increase the flow rate of the machine, three rotors (figure 2) or four rotors (figure 3) connected in parallel can be used.

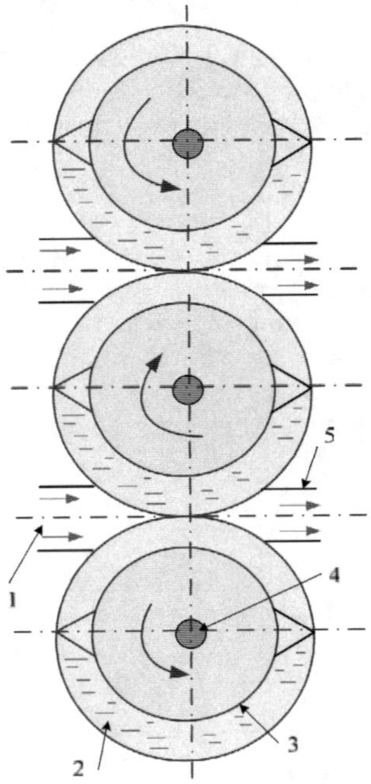

Figure 2. Rotating machine with three rotors. 1- suction chamber, 2- case, 3- rotor, 4- shaft, 5 - discharge chamber

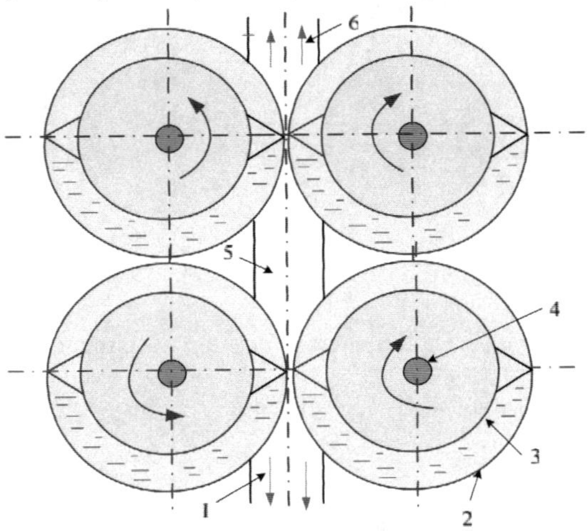

Figure 3. Rotating machine with four rotors. Rotating machine with three rotors. 1- suction chamber, 2- case, 3- rotor, 4- shaft, 5 – intermediary discharge chamber, 6- final discharge chamber

If the operation scheme from figure 2 is chosen, it means that the increase in the flow rate of the transported fluid is pursued; if the scheme in figure 3 is used, an increase in pressure will be obtained from p_1 to p_5 in the first stage and from p_5 to p_6 in the second stage. On the outside of the rotors in figures 2 and 3, gear wheels are fixed on the shafts that ensure the synchronous rotation of the rotors in the direction indicated in the figures.

The constructive solution does not contain elements that perform alternative rectilinear movements; it features safe operation and easy maintenance. In its construction, increased precision is required since if there are large gaps between the rotor and the case, the volumetric efficiency of the pump will decrease.

3. Results

The original constructive solution presented in this paper must specify the shape of the rotor contour.

The mathematical model for calculating the rotor profile contour coordinates has a general character; it can be modified for other dimensions of the rotor, it being necessary to specify only the values: z and R_r. After obtaining these coordinates, a program can be developed for the construction of the rotor on a numerically controlled machining center.

The following figure shows the execution drawings for the main component elements of the new type of rotating pump with original shape rotors.

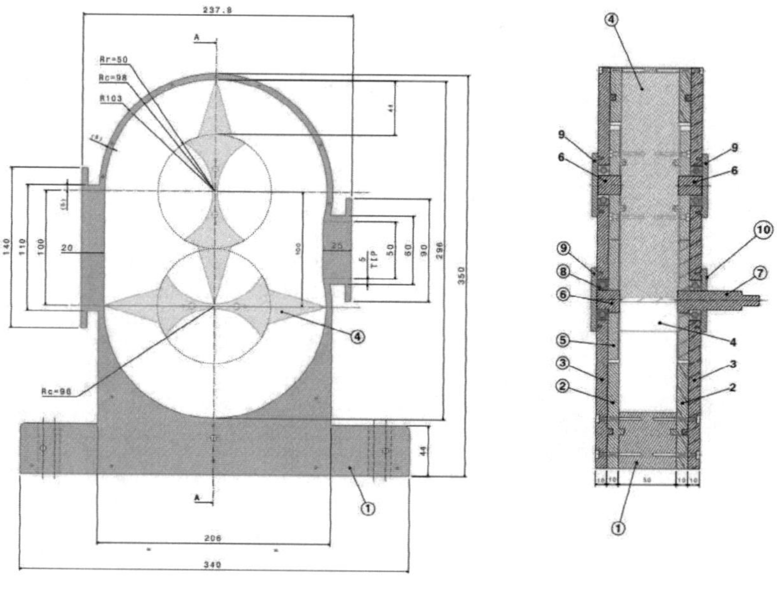

Figure 4. Cross section (a) and longitudinal section (b) through the rotating pump. 1 - oval case; 2 - intermediate wall; 3 - left side wall; 4 - rotating piston; 5 - gears; 6 - driven shaft; 7 - driving shaft; 8 - ball bearing; 9 - bearing cap; 10 - drive shaft bearing cap.

After running the program developed for the construction of the rotor on a numerically controlled machining center, the rotors in figure 5 resulted:

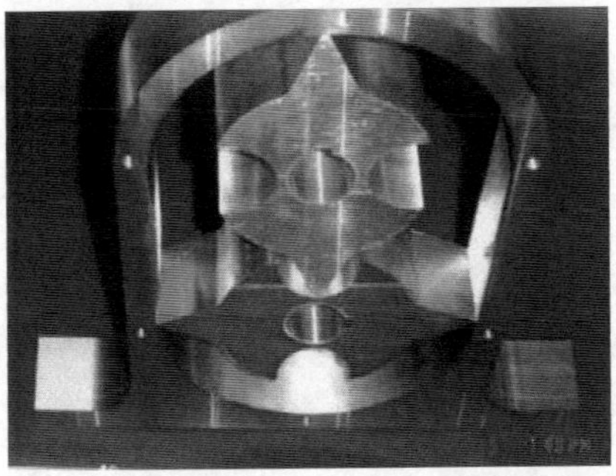

Figure 5. Axonometric view of the two rotors with triangular pistons mounted inside the case.

To calculate the flow rate, the following characteristic dimensions of the rotating volumetric pump must be specified: l = 0.05 [m]; z = 0.048 [m]; R_r = 0.05 [m]; n_r =100÷500 [rot/min].

Obviously, the following must be specified:

– The radius of the pump cylindrical casing (R_c): R_c = 0.05+0.048 = 0.098 m

– The speed of the pump will be changed by changing the speed of the three-phase asynchronous motor that drives it. The speed of the pump is limited by the value of the speed of the fluid at the suction of the pump, thus avoiding the phenomenon of discontinuity of the fluid (air voids) [14].

In the specialized literature [15] it is recommended that the velocity of the water in the suction line to the rotating machine is about 1 [m/s]. As a result, the pump speed will be within the limits of 100 ÷ 500 [rot/min]

In order to carry out the research, it is necessary that the speed of the pump can be changed; so, it appears necessary to change the speed of the electric motor driving the pump.

The volumetric flow rate variation at different speeds results and can be seen in figure 6.

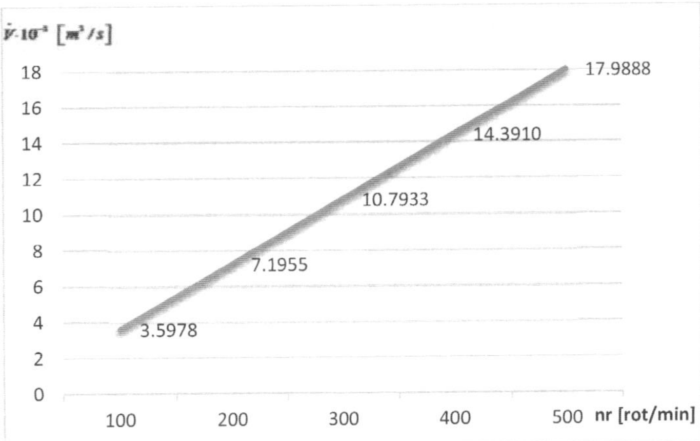

Figure 6. The volumetric flow rate variation at different speeds.

Based on the theoretical data one can obtain the driving power for the new constructive volumetric pump:

$$P = \dot{V}_m \cdot \Delta p = \left[\pi \cdot l \cdot z \cdot (z + 2R_r) \cdot \frac{n_r}{30} - V_p \right] \cdot \Delta p \; [W] \tag{6}$$

The total pressure increase achieved by the rotating machine (Δp) changes when the pump speed increases; the hydrostatic load and pressure losses occurring in the hydraulic circuit of the pump are evaluated at about H = 4 mH_2O [16]:

$$\Delta p = \rho_{H_2O} \cdot g \cdot H = 10^3 \cdot 9.81 \cdot 4 = 0.3924 \cdot 10^5 \ [Pa] \tag{7}$$

$$P = 0.0179888 \cdot 0.3924 \cdot 10^5 = 705.881 \, [W] \tag{8}$$

The results from theoretical calculations obtained are observed in figure 7.

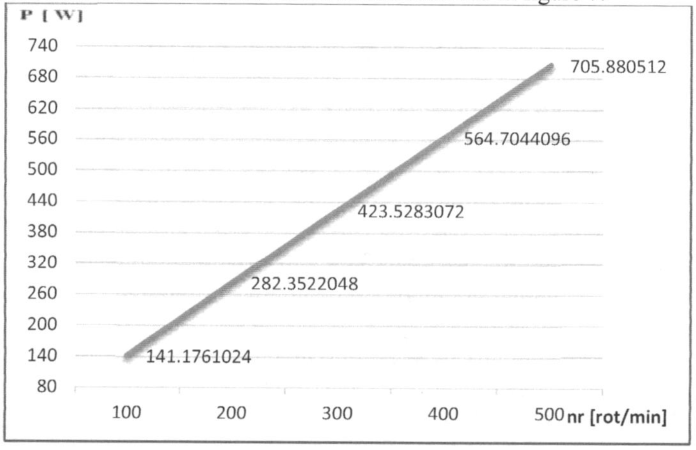

Figure 7. The driving power in function of the speed.

And the variation of the power in function of the speed is shown in figure 7.

4. Discussion

The actual driving power of the rotating volumetric pump will be higher:

$$P_r = \frac{P}{\eta_e} \ [W] \tag{9}$$

where η_e is the effective efficiency of the pump [12]:

$$\eta_e = \eta_v \cdot \eta_m \cdot \eta_h \tag{10}$$

where: η_v =0.9 is the volumetric efficiency of the pump; η_m = 0.8 is the mechanical efficiency of the pump; η_h =0.72 is the hydraulic efficiency of the pump. For the volumetric pump, in paper [12] the effective efficiency of the pump was determined, η_e = 0.8; this value was determined by tests in the Department of Thermotechnics, Engines, Thermal and Refrigerating Equipment of the Faculty of Mechanical and Mechatronics Engineering. So, for the volumetric pump: η_e =0.8.

5. Conclusions

Researchers are concerned about finding new application and use ways of this new type of working machine which can circulate any fluid substance.

The machine driving power will depend on the fluid nature, air, water, oil.

This paper is a contribution to the theoretical and experimental research in the field of rotating machine with application to the study of profiled rotors.

It can be concluded that the volumetric pump with profiled rotors is more advantageous than other pump types because it leads to energy saving for its driving [17].

Acknowledgements
The study has been funded by the Academy of Romanian Scientists within the National Projects AOSR-TEAMS II 2023-2024.

References
[1] Yang H.-Y., Pan M., 2015, Engineering research in fluid power: a review, J. Zhejiang. Univ-Sci. A. 16 (6), pp. 427–442.
[2] Alle N., Hiremath S.S., Makaram S., Subramaniam K., Talukdar A., 2016, Review on electro hydrostatic actuator for flight control, *Int. J. Fluid Power*, 17 (2), pp. 125–145. https://doi.org/10.1080/14399776.2016.1169743.
[3] Ivantysn J., Ivantysynova M., 2001, *Hydrostatic Pumps and Motor, Principles, Designs, Performance, Modeling, Analysis, Control and Testing*, Academic Books International, New Delhi.
[4] Darbani A.A., Shang L.Z., Beale J.R., Ivantysynova M., 2019, Slipper Surface Geometry Optimization of the Slipper/Swashplate Interface of Swashplate-Type Axial Piston Machines, *Int. J. Fluid Power*, 20 (2), pp. 245–270. https://doi.org/10.13052/ijfp1439-9776.2025.
[5] Wang S., 2012, Improving the volumetric efficiency of the axial piston pump, *J. Mech. Des*., 134 (11). https://doi.org/10.1115/1.4007361111001.
[6] Ivanovic L., Stojanovic B., Blagojevic J., Bogdanovic G., Marinkovic A., 2017, Analysis of the flow rate and the volumetric efficiency of the trochoidal pump by application of Taguchi method, *Thh. Vjesn*. 24, pp. 265-270. https://doi.org/10.17559/TV-20150429090420.
[7] Xu B., Hu M., Zhang J.H., 2016, Characteristics of volumetric losses and efficiency of axial piston pump with respect to displacement conditions, *J. Zhejiang*. Univ-Sci. A., 17, pp. 186–201. https://doi.org/10.1631/jzus.a1500197.
[8] Manring N.D., 2000, The discharge flow ripple of an axial-piston swash-plate type hydrostatic pump, J. *Dyn. Syst. Meas. Control-Trans*., ASME, 122, pp. 263–268. https://doi.org/10.1115/1.482452.
[9] Achten P.A.J., 2004, Power Density of the Floating Cup Axial Piston Principle, Proceedings of ASME 2004, Anaheim, California, USA, November 13–19, 11-22. https://doi.org/10.1115/IMECE2004-59006.
[10] Ericson L., Forssell J., 2018, A novel axial piston pump/motor principle with floating pistons: design and testing, Proceedings of the BATH/ASME 2018, UK. September 12-14. https://doi.org/10.1115/FPMC2018-8937.
[11] Băran N., Fischer G.S., Stoican (Prisecaru) M. M., 2021, A new solution for the transport of fluids in construction-related installations", *Asian Journal of Applied Science and Technology* (AJAST), 5, (2), pp. 51- 57.
[12] Stoican (Prisecaru) M. M., Băran N., 2020, A constructive solution that can function as a force machine or as a work machine, *Asian Journal of Applied Science and Technology* (AJAST), 4 (2), pp. 97-107.
[13] TcacencoV., 2005, *Machining centers with vertical axis – Alzmetall* (in Romanian), Technical and Technological Magazine, no. 4, Bucharest.
[14] Băran N., Zaid A., Besnea D., Duminica D., 2006, Theoretic and experimental research on a new type of rotating volumetric pump, *Romanian Review Precision Mechanics, Optics & Mecatronics,* 29, pp. 581-584.
[15] Ionescu M., 2010, *General hydraulics*, University of Petrol - Gases of Ploiesti.
[16] Bansal K., 2005, Fluid Mechanics and Hydraulic Machines, Delhi, India: Delhi College.
[17] Hawas M., Băran N., Detzortzis A., 2014, Influence of the rotor architecture and of the speed on the volumetric efficiency of a new type of rotating volumetric machine, *Advanced Materials Research*, Trans Tech Publications, Switzerland, vol. 905, pp. 487-491.

IOP Conf. Series: Materials Science and Engineering 1290 (2023) 012007 doi:10.1088/1757-899X/1290/1/012007

Identification of thermal and strength phenomena in power installations

P Duda and M Konieczny

Institute of Thermal and Process Engineering, Faculty of Mechanical Engineering, Cracow University of Technology, Al. Jana Pawła II 37, 31-864 Kraków, Poland

piotr.duda@pk.edu.pl

Abstract. The aim of this work is to show the inverse method, which can be used to determine the transient stress distribution caused by pressure and a variable temperature field. The method is fast enough to be used online in systems monitoring the thermal and strength condition of selected elements. The solution is possible despite the lack of information about the thermal boundary condition on the inner surface of the tank. The lack of data on the phenomena occurring inside the collector may be caused by the inability to install measuring sensors. Computational Fluid Dynamics (CFD) can take a long time to analyze complex phenomena on the inner surface of a collector, making it difficult to apply the method online. Solving the inverse problem is possible thanks to the measured temperature transients at selected points inside or on the outside of the component. After identifying the temperature distribution, stresses are determined using the finite element method. Then the maximum stresses can be compared with the permissible values, which allows the assessment of the operational safety of the element.

1. Introduction

The behavior of components of conventional, nuclear power plants or fusion power industry has been studied in numerous research projects. During their operation, high stresses occur in thick-walled pressure elements. Cyclical work caused by cooling and heating processes can cause the phenomenon of low-cycle fatigue and even lead to fatigue damage [1]. An old forced circulation dual-fuel subcritical power boiler was analyzed to reduce cyclic stress [2]. Thermal stresses in the tube walls were calculated based on the assumption of the linear temperature distribution in the tubes. Experimental and numerical studies of the safety of water walls in an opposed firing power boiler are presented in [3]. In the case of the proposed arc-shaped structure of the water wall, the thermal stresses can be significantly reduced. Pastor et al. [4] investigated the cause of the failure of a steam power boiler, in which numerous cracks were found after its shutdown. The cause turned out to be a failure of the control unit, which led to a sudden temperature change in the piping system. As a result, the stresses exceeded the permissible values, which led to the destruction of the welded joint. The analyzes carried out in the cited works required the construction of a geometric model, the specification of material properties, and the definition of all boundary conditions. If any of these data cannot be defined, an inverse problem is formulated instead of a direct problem. Oliveira et al.[5] investigated the accuracy of the linear inverse model for the estimation of the boundary heat flux with a material undergoing a phase change. The solution was possible despite the lack of the boundary condition thanks to the temperature transients

Content from this work may be used under the terms of the Creative Commons Attribution 3.0 licence. Any further distribution of this work must maintain attribution to the author(s) and the title of the work, journal citation and DOI.
Published under licence by IOP Publishing Ltd

that were generated from the direct solution. The linear inverse model was positively verified when estimating the limit heat flux if the specific heat did not strongly depend on the temperature. Identification of thermal boundary conditions on the basis of measured temperature transients in a power boiler is shown in [6]. The analyzed processes did not change quickly over time, so it was possible to assume a steady state at a given time. The calculations were made iteratively using the conjugate gradient method. Experimental and numerical studies on concentrated solar power systems are presented in [7]. A loop of molten salt was placed on the test bench to model the receiver tube of the solar tower. Based on the measured temperature distribution on the outer surface, the incident heat flux onto the receiver tube was identified. For the inverse unsteady solution, the control volume method in two dimensions and cylindrical coordinates was used. Temperature-independent material properties were assumed and calculations were carried out iteratively. The method to solve the three-dimensional inverse problem in a simple-shaped region is shown in [8]. It is much more difficult to solve inverse problems in areas with complex shapes and temperature-dependent material properties [9-10].
The aim of this work is to show the inverse method, which can be used to determine the transient stress distribution caused by pressure and a variable temperature field. The method is fast enough to be used online in systems monitoring the thermal and strength condition of selected elements. The solution is possible despite the lack of information about the thermal boundary condition on the inner surface of the tank. The lack of data on the phenomena occurring inside the collector may be caused by the inability to install measuring sensors. Computational Fluid Dynamics (CFD) can take a long time to analyze complex phenomena on the inner surface of a collector, making it difficult to apply the method online. Solving the inverse problem is possible thanks to the measured temperature transients at selected points inside or on the outside of the component. After identifying the temperature distribution, stresses are determined using the finite element method. Then the maximum stresses can be compared with the permissible values, which allows the assessment of the operational safety of the element.

2. Problem description

Transient physical phenomena occur in the thick-walled cylindrical collector shown in Fig. 1. It is assumed that the dimensions of the collector and the properties of the material from which it is made are known. In addition, it is assumed that the analyzed phenomenon is two-dimensional and symmetrical with respect to the vertical surface. Temperature measurement points are installed on the outer surface in half of the cross-section, thanks to which it will be possible to identify the unknown temperature and stress distribution. It is assumed that the boundary condition on the outer surface is known.

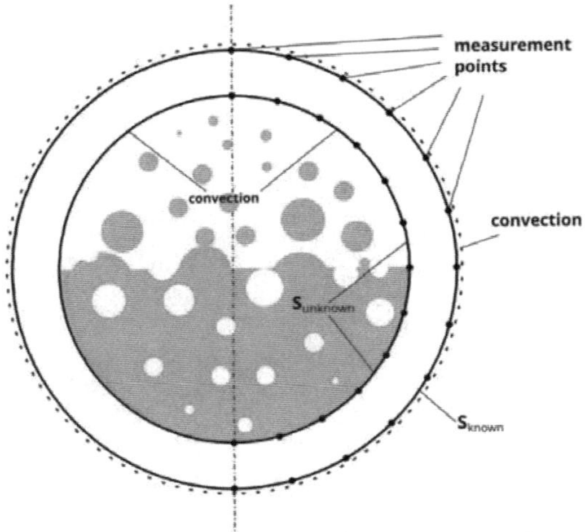

Figure 1. Cylindrical collector cross-section.

IOP Conf. Series: Materials Science and Engineering 1290 (2023) 012007 doi:10.1088/1757-899X/1290/1/012007

The transient heat conduction equation for an isotropic and homogeneous body without heat sources has the following form

$$c(T)\rho(T)\frac{\partial T}{\partial t} = \nabla \cdot \left(k(T)\nabla T\right) \tag{1}$$

Time constants of the temperature sensors used do not allow the registration of very fast temperature changes as analyzed in [11]. For this reason, the parabolic heat conduction equation was used, not a hyperbolic one. Non-uniform temperature distribution causes non-uniform material displacement, generates strains and thermal stresses in components. It is assumed that stresses should not exceed the allowable values, because this is the aim of a monitoring algorithm. Even maximum deformations should not be a source of heat that should be involved in thermal analysis. For this reason, it is assumed one way coupling between thermal and mechanical analyses. The resulting transient temperature distribution is taken for mechanical analysis.
The following thermoelasticity equations can be used for structural analysis

$$\sigma_{ij,j} + f_i = 0 \tag{2}$$

$$\varepsilon_{ij} = \frac{1}{2}\left(u_{i,j} + u_{j,i}\right) \tag{3}$$

$$\sigma_{ij} = 2G\varepsilon_{ij} + \lambda\varepsilon_{kk}\delta_{ij} - \beta T\delta_{ij} \tag{4}$$

where tensor, vector or scalar variables mean respectively: σ - stress, **f** - body force, **u** - displacement, ε - strain, δ_{ij} –Kronecker delta c – specific heat, ρ – density, k – thermal conductivity, G, λ – Lame constants, β– thermal expansion coefficient.
The analyzed domain was divided into two-dimensional finite volumes during inverse thermal calculations and into three-dimensional finite elements during strength calculations, as shown in Figure 2.

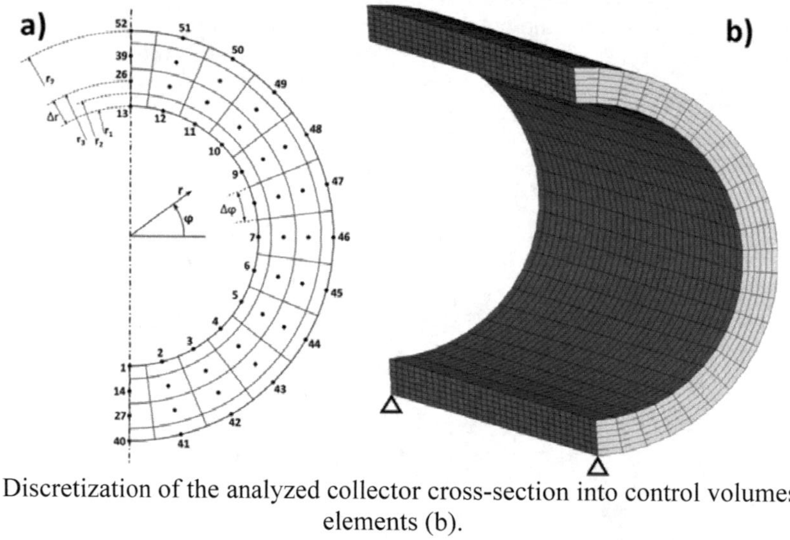

Figure 2. Discretization of the analyzed collector cross-section into control volumes (a) finite elements (b).

IOP Conf. Series: Materials Science and Engineering 1290 (2023) 012007 doi:10.1088/1757-899X/1290/1/012007

The inverse calculation starts with a surface with a known boundary condition, on which thermocouples are additionally placed. The heat balance equation for the control volume associated with node i=40, which is presented in figure 2a can be formulated as

$$\frac{dT_i}{dt}=\frac{4}{c_i\rho_i\left(r_7^2-r_6^2\right)\cdot\Delta\varphi}\left[\frac{k_{i+1}+k_i}{2}\cdot\frac{0.5\cdot\Delta r}{r_7\cdot\Delta\varphi}\left(T_{i+1}-T_i\right)+\frac{k_{i-13}+k_i}{2}\cdot\frac{r_6\cdot 0.5\cdot\Delta\varphi}{\Delta r}\left(T_{i-13}-T_i\right)\right] \tag{5}$$

where $i=40,\quad k_i=k\left(T_i\right),\quad c_i=c\left(T_i\right),\quad c_i=c\left(T_i\right),\quad \rho_i=\rho\left(T_i\right)$.

The heat balance equations for control volumes i=41–51 have the following form

$$\frac{dT_i}{dt}=\frac{2}{c_i\rho_i\left(r_7^2-r_6^2\right)\cdot\Delta\varphi}\left[\frac{k_{i-1}+k_i}{2}\cdot\frac{0.5\cdot\Delta r}{r_7\cdot\Delta\varphi}\left(T_{i-1}-T_i\right)+\frac{k_{i-13}+k_i}{2}\cdot\frac{r_6\cdot\Delta\varphi}{\Delta r}\left(T_{i-13}-T_i\right)+\right.$$
$$\left.+\frac{k_{i+1}+k_i}{2}\cdot\frac{0.5\cdot\Delta r}{r_7\cdot\Delta\varphi}\left(T_{i+1}-T_i\right)\right] \tag{6}$$

The equation for node i =52 is expressed as:

$$\frac{dT_i}{dt}=\frac{4}{c_i\rho_i\left(r_7^2-r_6^2\right)\cdot\Delta\varphi}\left[\frac{k_{i-1}+k_i}{2}\cdot\frac{0.5\cdot\Delta r}{r_7\cdot\Delta\varphi}\left(T_{i-1}-T_i\right)+\frac{k_{i-13}+k_i}{2}\cdot\frac{r_6\cdot 0.5\cdot\Delta\varphi}{\Delta r}\left(T_{i-13}-T_i\right)\right] \tag{7}$$

The temperature transients in nodes i =40–52 are measured and temperature derivatives versus time can be calculated in these nodes. If the material properties c, ρ and k are temperature-independent, temperatures $T_{27}-T_{39}$ can be calculated form equations (6–8). If the material properties are temperature-dependent, the equations may be solved iteratively. In a similar way, temperatures $T_{14}-T_{26}$ and T_1-T_{13} can be determined.
Compared to direct problems, inverse problems are more difficult to solve because they are ill-posed. To stabilize the solution, smoothing filters are applied. A third-order polynomial with eleven subsequent measurement points is used to smooth out the measured and calculated temperature transients [12].
Unlike thermal analysis, strength simulation is not two-dimensional but three-dimensional. Half of the collector was divided into: 30 finite elements along the axis, 25 along the circumference and 6 along the wall thickness. The model is constrained to prevent rigid motion and to meet symmetry conditions. A time-varying pressure was assumed on the inner surface and appropriate longitudinal stresses were introduced. Finally, an unsteady temperature distribution was introduced, which was determined in the thermal analysis.

3. Numerical verification

The inverse method accuracy can be estimated in the numerical test for the collector, which is shown in figure 3. It is made of 18GMNA steel, has an outer diameter of 1.8 m and a thickness of 115 mm. The collector with the initial uniform temperature of T_0 = 100°C is flooded by the medium. The changes of fluid temperature and pressure over time are presented in figure 4. At the beginning of the heating process, the heating rate is 5 K/min, later 2.5 K/min, and at the end it is 5.7 K/min. The pressure changes according to the saturation parameters for water. The heat transfer coefficient on the inner surface is assumed to be constant in time but varies depending on the angle. It takes the highest value of 3000 W/m^{2}K at the top point and the lowest value of 300W/m^{2}K at the bottom point. The outer surface is in contact with the air (T_{out} = 20°C) and the heat transfer coefficient is estimated to be 5 W/m^{2}K.

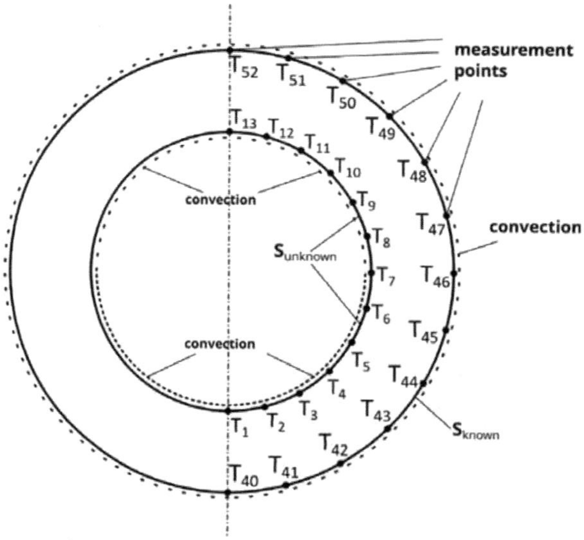

Figure 3. Assumed temperature measurement points for the analysed collector.

For all given boundary conditions, initial conditions, geometry and material properties, the problem of transient heat conduction can be solved by the direct method by finding the unsteady temperature distribution including transients at 13 points on the outer surface. For this purpose, the ANSYS program [13] was used, which is based on the finite element method. A denser division was applied than during the inverse solution, by dividing half of the cross-section into 15 plane four-node finite elements along the thickness and 120 along the circumference. The temperature transients calculated by the direct method were then disturbed by random errors of ± 0.25°C with the average value equal to zero and they are presented in figure 4.

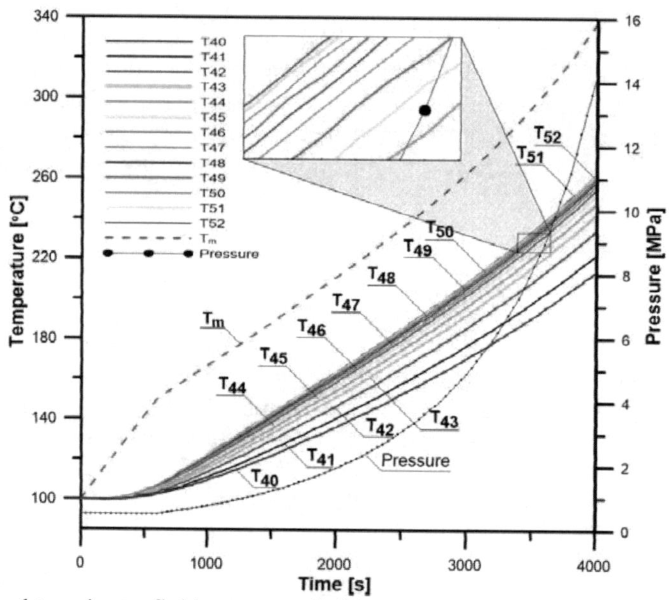

Figure 4. Measured transients: fluid pressure, fluid temperature and temperature in nodes 40–52.

Then, it is assumed that the boundary condition on the inner surface of the component is unknown. This situation occurs in real installations. The variable velocity and temperature of the medium, as well as the occurring phase changes, make it difficult to estimate this boundary condition. Measurement of temperature and heat flux on the inner surface is also difficult due to high operating parameters. If temperature transients on the outer surface with a known boundary condition are available, the problem can be solved using the inverse method.

At first, equations (5-7) are used to determine the temperature transients in the volumes assigned to nodes 27-39. In a similar way, temperatures are determined in volumes related to nodes in subsequent layers, i.e. 14-26 and 1-13. The calculated temperature transients in volumes corresponding to nodes 1-13 are shown in Fig. 5. Despite the distortion of the measurement data by randomly generated errors, the identified temperature transients are very similar to those obtained by the direct method , as can be seen in Fig. 5.

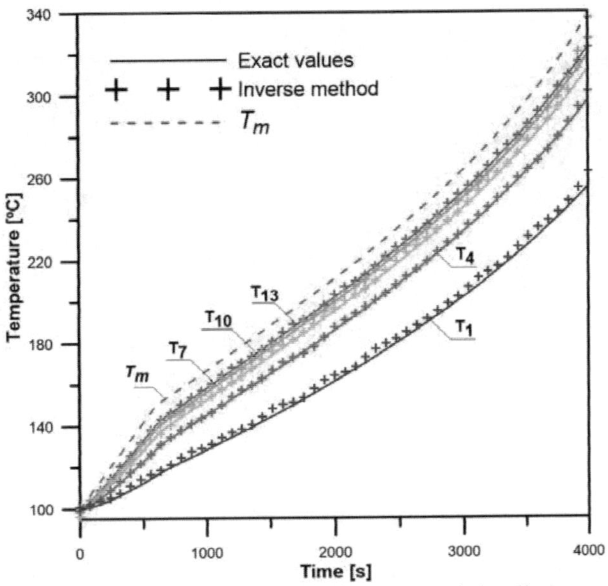

Figure 5. Temperature transients obtained from the inverse and the direct method on the collector inner surface.

In solving inverse problems, there are often significant oscillations that do not occur under real conditions. They are caused by poor conditioning of the problem due to the lack of one of boundary conditions. Even small errors in temperature measurements can generate significant errors in identified temperature fields. Various procedures have been proposed to stabilize the inverse solution, including regularization, future steps and digital smoothing filters [12]. The best results for time smoothing are obtained using the third-order eleven-point polynomial filter. The same filter is used for the calculation of the derivative *dT/dt*. The solution stability depends also on the time step Δt. Choosing the time step for the adopted smoothing filter can be done iteratively, starting from the smallest step and gradually increasing it. The presented results are obtained using the time step of $\Delta t = 40$ s.

Accurate determination of the temperature distribution is crucial for the thermal stress estimation. The most important is the information on maximum total stresses and the phenomenon duration. This knowledge can be used in heating process optimization and lifespan prediction.

The radial stresses arising on the inner surface during the heating process are shown in figure 6. Transients determined by the inverse method are similar to those determined by the direct method(exact values). Compared to radial stresses, the magnitude of circumferential and longitudinal stresses is one order higher (see figures 7–9). Compressive stresses reach the highest values on the inner surface of the component upper part, where the highest heat transfer coefficient appears. The maximum value of circumferential stresses after 680 s of the heating operation reaches 80 MPa, as can be seen in figure 7. The longitudinal stresses on the inner layers are presented in figure 8. The highest compressive stress on the inner surface does not occur in the highest point but in node 7 and totals 103 MPa. When the inner surface is affected by compressive stresses, tensile stresses appear on the outer surface, as can be seen in figure 9. The highest values of tensile stresses are observed on the outer surface in node 49.

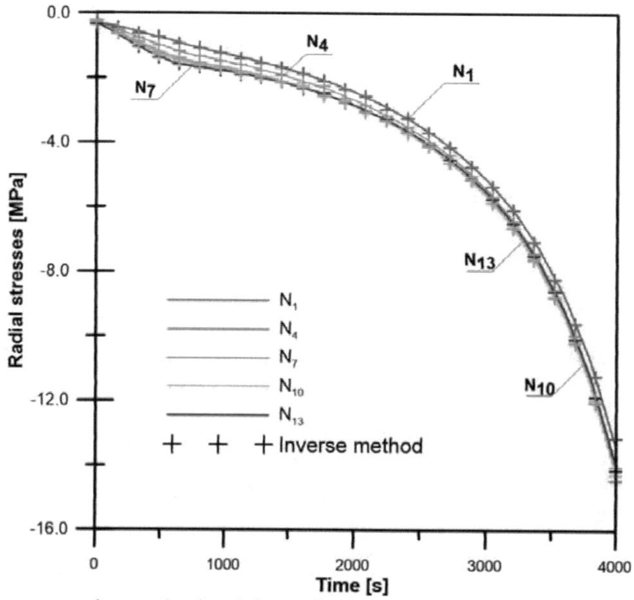

Figure 6. Radial stress transients obtained from the inverse and the direct method on the collector inner surface.

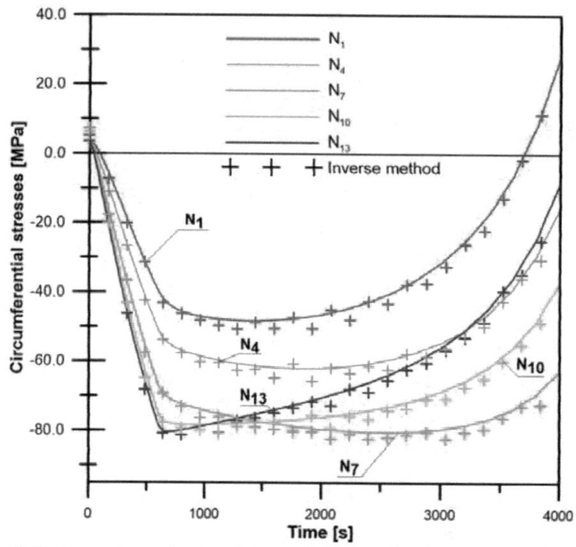

Figure 7. Circumferential stress transients obtained from the inverse and the direct method on the collector inner surface.

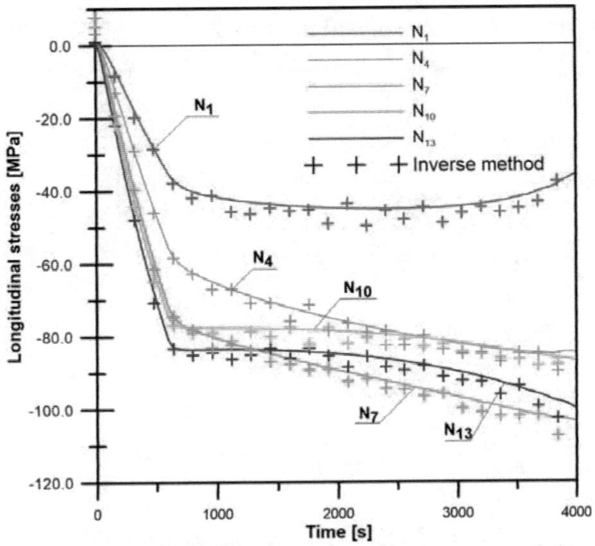

Figure 8. Longitudinal stress transients obtained from the inverse and the direct method on the collector inner surface.

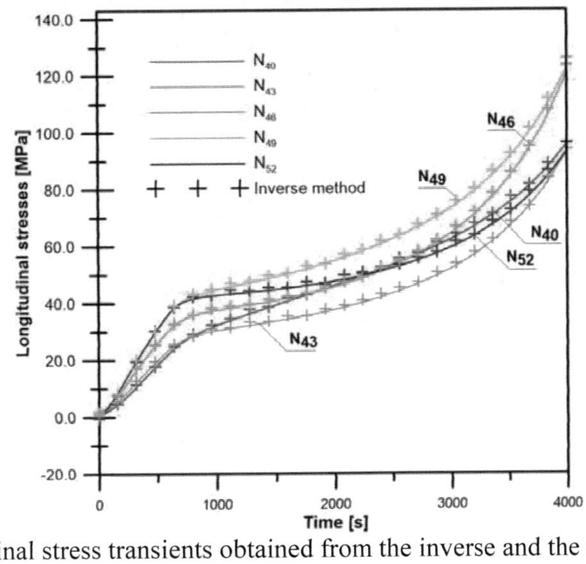

Figure 9. Longitudinal stress transients obtained from the inverse and the direct method on the collector outer surface.

Finally, the Huber-Hanky-Mises hypothesis is applied to find equivalent stresses and they are shown in figure 10 and figure 11. Like the stresses shown earlier, they are caused by the uneven temperature field and fluid pressure. In addition, they take into account the complex state of stress. Comparing the equivalent stresses to the stress components presented earlier, it can be seen that they are mainly caused by circumferential and longitudinal stresses. The highest values appear at the top of the inner surface because that's where the heat transfer coefficient is highest. The highest equivalent stress in the first part of the process is 81 MPa and occurs after 680 s in node 13, which is shown in figure 3. Even greater stresses appear at the end of the heating process when the heating rate is the highest.
The relative error of the determined equivalent stress calculated on the basis of comparison with the direct solution is 3.5%. Figure 11 shows the equivalent total stress distribution after 760 s of the heating

process. The maximum values appear at the inner surface, the smallest in the middle of the wall thickness.

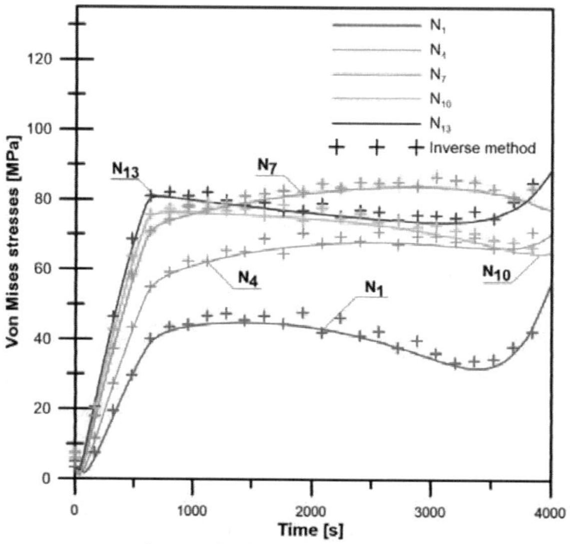

Figure 10. Equivalent stress transients obtained from the inverse and the direct method on the collector inner surface.

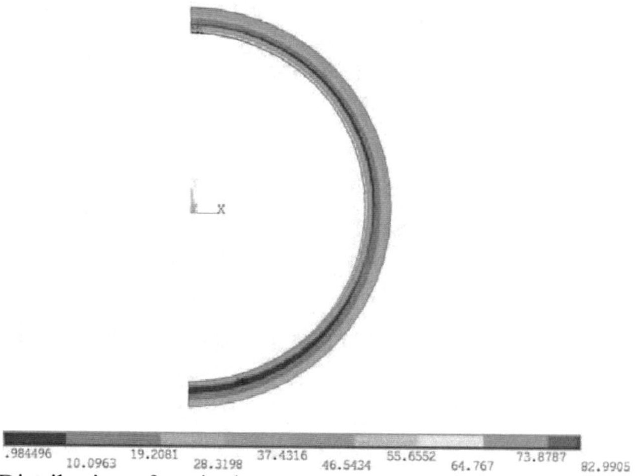

Figure 11. Distribution of equivalent stress after 760 s of the heating process.

4. Summary and conclusions

The paper shows an inverse method, which can be used to determine the transient stress distribution caused by pressure and a variable temperature field. The method is fast enough to be used online in systems monitoring the thermal and strength condition of selected elements. The solution is possible despite the lack of information about the thermal boundary condition on the inner surface of the tank. After identifying the temperature distribution, stresses are determined using the finite element method. The performed numerical tests show the stability and accuracy of the proposed inverse method based on the identified temperature and total stress transients. The relative error of the determined equivalent stress calculated on the basis of comparison with the direct solution is 3.5%. The identified radial, circumferential, longitudinal and equivalent stress transients are shown. The results indicate that the highest stress does not exceed the allowable values (–198.567, 511.310 MPa) which were calculated on

IOP Conf. Series: Materials Science and Engineering 1290 (2023) 012007 doi:10.1088/1757-899X/1290/1/012007

the basis of TRD 301 regulations [14]. However, the obtained strain fields are large enough to record them using strain gauges. The method can be particularly useful where it is not possible to drill holes for sensors in the walls of pressure elements.

References

[1] Vetri Selvan R. et al. 2019 The effect of ligament size on the thermal fatigue life of 9Cre1Mo steel boiler header under cold, warm and hot starts. *Engineering Failure Analysis*, 97, 727–739.

[2] Mahmoudi F. et al. 2022 Developing an operating instruction to attenuate pressure stress during fuel shortage events of a power boiler based on dynamic modelling. *Applied Thermal Engineering*, 200 117565.

[3] Wei L. et al. 2022 Experimental study and simulation analysis of heat and deformation in the water walls of an opposed firing boiler under flexible operating conditions. *Applied Thermal Engineering* 213 118726.

[4] Pástor M. et al. 2020 Prediction of failures in steam boiler using quantification of residual stresses. *Engineering Failure Analysis* 118 104808.

[5] Oliveira A.V.S. et al. 2023 Using a linear inverse heat conduction model to estimate the boundary heat flux with a material undergoing phase transformation. *Applied Thermal Engineering* 219 119406.

[6] Taler J, Duda P, Węglowski B, Zima W, Grądziel S, Sobota T and Taler D 2009 Identification of local heat flux to membrane water-walls in steam boilers. *Fuel*, *88*(2), 305-311.

[7] Fernández-Torrijos M. et al. 2019 Inverse heat problem of determining unknown surface heat flux in a molten salt loop, *International Journal of Heat and Mass Transfer*, 139 503–516.

[8] Alifanov O M and Nenarokomov A V 1999 Three-dimensional boundary inverse heat conduction problem for regular coordinate systems. *Inverse problems in engineering*, *7*(4), 335-362.

[9] Duda P 2015 Solution of an inverse axisymmetric heat conduction problem in complicated geometry. *International Journal of Heat and Mass Transfer*, *82*, 419-428.

[10] Duda P 2016 A method for transient thermal load estimation and its application to identification of aerodynamic heating on atmospheric reentry capsule. *Aerospace science and technology*, *51*, 26-33.

[11] Majchrzak E and Mochnacki B 2018 Dual-phase lag model of thermal processes in a multi-layered microdomain subjected to a strong laser pulse using the implicit scheme of FDM. *International Journal of Thermal Sciences*, *133*, 240-251.

[12] Duda P and Konieczny M 2019 A new algorithm for solving an inverse transient heat conduction problem by dividing a complex domain into parts. *International Journal of Heat and Mass Transfer*, *128*, 865-874.

[13] ANSYS User's Manual, Revision 15.0A.

[14] TRD 301, 1986 Technische Regeln für Dampfkessel. Carl Heymans Verlag, Köln und Beuth-Verlag, Berlin, 98-138

Exergy analysis, coupled heat pump and refrigeration system using waste heat

D Dima[1] and A Dobrovicescu[1*]

[1]Faculty of Mechanical Engineering and Mechatronics, University Politehnica of Bucharest

E-mail: adobrovicescu@yahoo.com

Abstract. The paper seeks the optimal configuration of equipment capable of generating heat and cold. For the generation of the two products, for the first time it is considered a coupled system heat pump and refrigeration system. Waste heat from an industrial process is used as the heat source for the heat pump cycle. For the refrigeration system, carbon dioxide is used as the working agent, and for the heat pump the working agent R152a. In order to identify the malfunctions at the level of each device and process, an exergy analysis is used, which identifies the magnitude and location of each malfunction. Based on the information provided by the exergy analysis, it is considered a structural optimization strategy by applying sub-cooling to the refrigeration cycle and heat pump cycle. The analysis shows an increase in the performance indices COP_{en} and COP_{ex} when applying the subcooling at the level of the refrigeration cycle. The results of the exergy analysis make it possible to find solutions for improving the functional structure and design of equipment.

Keywords: exergy analysis, exergy destruction, heat pump, waste heat.

1. Introduction

Today, global warming is a fundamental problem. The implementation of environmentally friendly technologies regarding the heating of residential spaces and the preparation of domestic hot water is a priority.

The COP26 [1] Glasgow Climate change Conference presented the program of reducing and phasing out fossil fuel consumption and replacing technologies that generate greenhouse gases. COP26 decisions will be successful if it includes heating and domestic hot water preparation with heat pump equipment. These technologies have evolved greatly over the last decade, being applicable to both domestic and industrial applications.

Heat pumps are technologies that do not generate polluting emissions with negative potential to the environment, other than those emitted by the production of electric energy. If the electricity consumed by the heat pump compressor comes from environmentally friendly sources such as solar or wind, the positive effect of these equipment on greenhouse gas reduction can be quantified.

Due to the greenhouse gas reduction policy [2] research is focused on heat pump equipment that uses natural working agents that are part of the natural biosphere. Carbon dioxide is a promising working

Content from this work may be used under the terms of the Creative Commons Attribution 3.0 licence. Any further distribution of this work must maintain attribution to the author(s) and the title of the work, journal citation and DOI.
Published under licence by IOP Publishing Ltd

agent with a wide range of use ensuring proper operation of the equipment even at large temperature differences between the hot source and the cold heat source [3-4].

Heat pumps have a substantial advantage having a superunit performance coefficient over classical heating equipment that have subunit thermal efficiency.

These equipment can integrate and recover heat low energy potential waste from industrial processes and contribute to the environmental greening. The recovery of heat from waste leads to a reduction in electricity consumption and increases the performance coefficient of the equipment.

Jian Li et al.[5] performs a thorough energy analysis of the carbon dioxide heat pump cycle by recovering waste heat. The coefficient of performance records a significant increase compared to the non-recovery heat pump cycle.

Zy XU et al.[6] in 2022 conducted an exergy analysis focused on the comparison of the heat pump cycle in a mechanical vapor compression stage and the absorption heat pump cycle, using the residual heat from industrial processes at the level of the cold source.

Baomin Dai et al.[7] investigates energetically and exergetically the most well-known heat pump cycles capable of delivering water at high temperatures using environmentally friendly working agents. The results of the study show that the greatest damage to exergy is recorded in the compression process.

Wei-wei et al.[8] conducts a theoretical analysis of equipment with internal heat exchanger, which can provide heat at temperatures above 80°C and cold at 7°C, using carbon dioxide as a working agent. Theoretical analysis shows that the discharge pressure at the level of the gas cooler has a great influence on the coefficient of performance.

This paper aims to investigate from an energy and exergy point of view a cycle of coupled heat pump that is capable of generating heat and cold, recovering the thermal utility of a waste water flow from an industrial process with a temperature above that of the environment.

Although many heat pump working agents are known in 2022, due to the global policy on decarbonization and greenhouse gas reduction, the attention of researchers is directed to environmental working agents. Some promising working agents by their environmental and energy characteristics are shown in Table 1.

Table 1. Ecological working agents used in heat pump cycles [9-10].

Working agents	Chemical formula	GWP	Critical temperature [°C]	Critical pressure [bar]
R744	CO2	1	30.98	73.77
R152A	C2FH4	124	113.26	45.20
R1234yf	C3F4H2	4	94.71	33.82
R1234ze (E)	C3F4H2	7	109.37	36.34

2. Description of the system

In order to satisfy the requirement of generating hot utility and cold utility but also of integrating and recovering the heat from the waste heat source, the equipment is imagined according to figure 1.

The refrigeration cycle 1 which ensures the cold utility at the level of the evaporator where the carbon dioxide is used as working agent and a heat pump cycle 2 which takes heat from the waste heat at the level of the evaporator using as working agent the working agent R152a.

Due to the large difference between the inlet temperature 11 of the heated water and the outlet temperature 13 of the heated water, the equipment is designed so that the water is heated in two stages. The first time the water is preheated by the gas cooler of the refrigeration installation to an intermediate temperature 12 and later the water is heated to the final temperature 13 by the condenser of the heat pump installation.

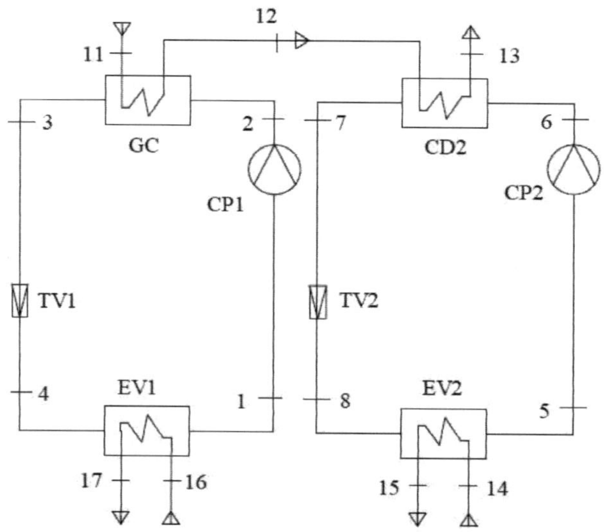

Figure 1. Schematic diagram of the equipment: 1) refrigeration cycle; 2) heat pump cycle

The main parameters that describe the equipment are specified in table 2.

Table 2. Main parameters describing the equipment.

Description	Status / measurement units	Value
Heated water inlet temperature	11 [°C]	20
Heated water outlet temperature	13 [°C]	100
Chilled water inlet temperature	16 [°C]	12
Chilled water outlet temperature	17 [°C]	7
Waste water inlet temperature	14 [°C]	50
Minimum temperature difference at Pinch in heat exchangers	ΔtP [°C]	5
Ambient temperature	T_0 [K]	293

3. Mathematical modeling

By setting the parameters presented in table 2, the mathematical model for evaluating the performance of the equipment can be developed.

The evaporator 1 ensures the cold utility, cooling the water from temperature 16 to temperature 17. The flow of working agent from the refrigeration cycle can be calculated:

$$\dot{m}_1 = \frac{\dot{Q}_{EV1}}{h_1 - h_4} \tag{1}$$

The heat pump cycle heats the water from the intermediate temperature 12 to the final temperature 13. The flow of working agent circulated in the heat pump cycle can be calculated:

$$\dot{m}_2 = \frac{\dot{Q}_{CD2}}{h_6 - h_7} \tag{2}$$

Mechanical work consumed by the refrigeration cycle:

$$\dot{W}_1 = \dot{m}_1 (h_1 - h_2) \tag{3}$$

Mechanical work consumed by the heat pump cycle:

$$\dot{W}_2 = \dot{m}_2(h_5 - h_6) \tag{4}$$

The energy performance coefficient of the global system can be calculated:

$$COP_{en} = \frac{\dot{Q}_{EV1}+\dot{Q}_{GC}+\dot{Q}_{CD2}}{\dot{W}_1+\dot{W}_2} \tag{5}$$

To identify the dysfunctions at the level of each device and process, exergy analysis is used. The exergy balance equation for a closed system:

$$\sum \dot{Ex}_Q^T = \sum \dot{W} + \sum \dot{I} \tag{6}$$

Applying equation (6) to the system described in figure 1, the exergy balance equation written at the level of the working agent is:

$$\left|\dot{Ex}_{Q_{EV1}}^{T_{EV1}}\right| + \left|\dot{Ex}_{Q_{EV2}}^{T_{EV2}}\right| + \left|\dot{Ex}_{Q_{GC}}^{T_{GC}}\right| + \left|\dot{Ex}_{Q_{CD2}}^{T_{CD2}}\right| = |\dot{W}_1| + |\dot{W}_2| + \dot{I}_{CP1} + \dot{I}_{CP2} + \dot{I}_{TV1} + \dot{I}_{TV2} \tag{7}$$

By identifying the fuel and products equation (7) can be rewritten at the level of the consumer of hot utility and cold utility.

$$|\dot{W}_1| + |\dot{W}_2| + \left|\dot{Ex}_{Q_{14-15}}^{T_{14-15}}\right| = \left|\dot{Ex}_{Q_{16-17}}^{T_{16-17}}\right| + \left|\dot{Ex}_{Q_{11-12}}^{T_{11-12}}\right| + \left|\dot{Ex}_{Q_{12-13}}^{T_{12-13}}\right| + \dot{I}_{CP1} + \dot{I}_{\Delta TGC} + \dot{I}_{TV1} + \dot{I}_{\Delta TEV1} + \dot{I}_{CP2} + \dot{I}_{\Delta TCD2} + \dot{I}_{TV2} + \dot{I}_{\Delta TEV2} \tag{8}$$

Based on the information provided by the heat exergy, the heat flow taken from the waste heat source has an exergy that is consumed by the system. For a correct balance this exergy must be mentioned as fuel.

Where "I" represents the destruction of exergy calculated on the basis of Gouy-Stodola's theorem [12-13].

$$\dot{I}_1 = T_0\dot{S}_{gen} \tag{9}$$

The share of exergy damage:

$$\Psi[\%] = \frac{\dot{I}_1}{\dot{W}_1+\dot{W}_1+\dot{Ex}_{Q_{14-15}}^{T_{14-15}}} 100 \tag{10}$$

Calculation of the exergy performance coefficient:

$$COP_{ex} = \frac{\dot{Ex}_{Q_{16-17}}^{T_{16-17}} + \dot{Ex}_{Q_{11-12}}^{T_{11-12}} + \dot{Ex}_{Q_{12-13}}^{T_{12-13}}}{\dot{W}_1+\dot{W}_2} \tag{11}$$

Although in the mathematical relation of the exergy balance (8) the exergy of heat taken from the waste heat source is fuel, its cost is zero and in the relation of the exergy performance coefficient (11) it does not appear.

4. Results and discussions

Thermodynamic modeling is performed based on the mathematical model described. The simulation of the operation and the investigation of the equipment performance was performed in the Engineering Equation Solvers work program [14].

IOP Conf. Series: Materials Science and Engineering 1290 (2023) 012008 doi:10.1088/1757-899X/1290/1/012008

4.1. Cycles connected on the hot water side

The decisional parameter is the pressure at the level of the gas cooler, of the refrigeration cycle.

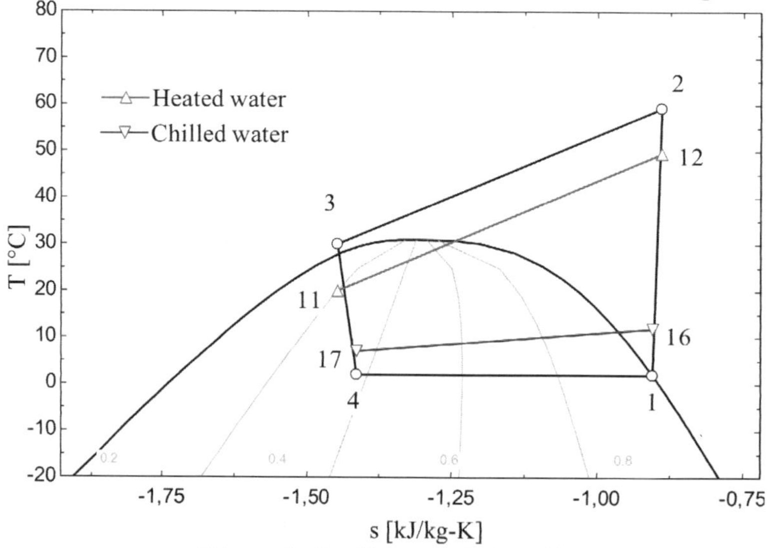

Figure 2. T-s diagram of the refrigeration cycle

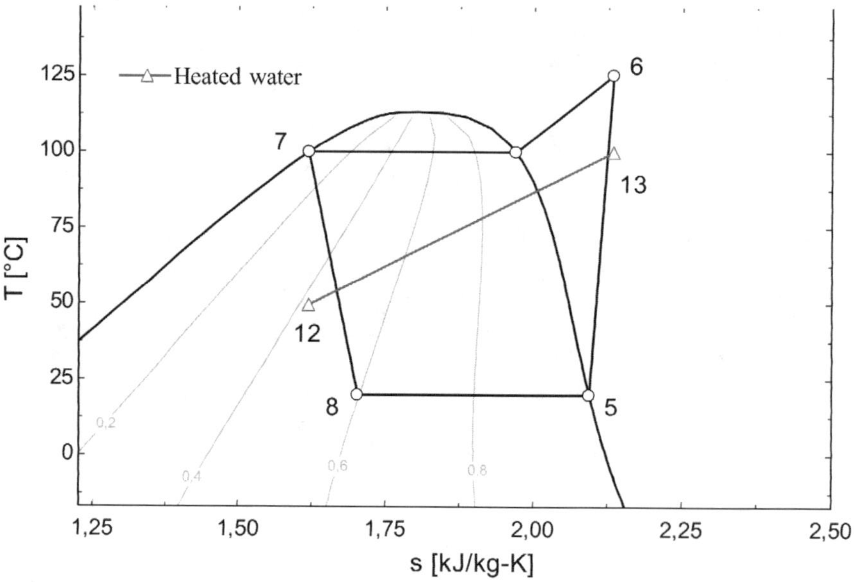

Figure 3. T-s diagram of the heat pump cycle

The diagrams in figure 2 and figure 3 show the evolution of the refrigeration cycle and the evolution of the heat pump cycle in entropy temperature coordinates.

Following figure 4 which indicates the exergy destruction at the level of each appliance and process at the variation of the discharge pressure of the refrigeration cycle it can be seen that the exergy destruction at the level of the throttling processes of the refrigeration cycle and the heat pump cycle registers the highest values.

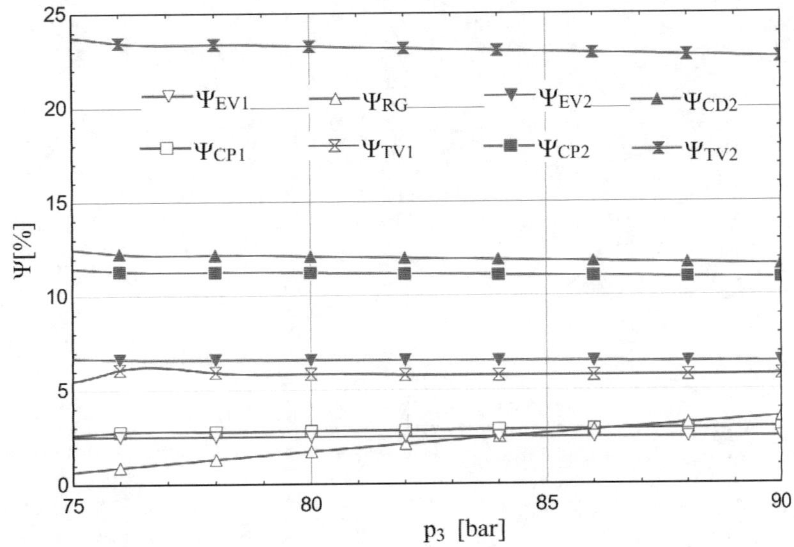

Figure 4. Exergy destruction in each device and process cycles

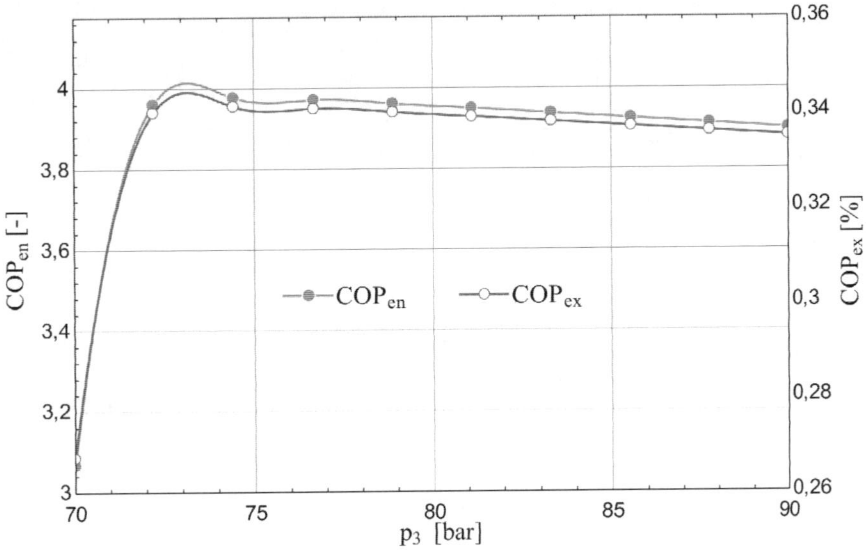

Figure 5. COP_{en} and COP_{ex} performance indices

According to figure 5, an optimal discharge pressure can be established at the level of the gas cooler, which can establish a maximum coefficient of performance.

Based on the information provided by the exergy analysis that identifies the location and magnitude of any defect, an optimization strategy can be considered in order to improve the exergy performance coefficient.

According to the diagram in figure 4, the throttling processes have priority in order to optimize, so in the case of the refrigeration cycle regenerative subcooling is applied and in the case of the heat pump cycle, subcooling can be achieved with heated water because the intermediate temperature 12 is significantly lower than condensation in the heat pump cycle.

4.2. Cycles connected on the hot water side and regenerative subcooling

Figure 5 shows the diagram of the equipment with regenerative subcooling in the refrigeration cycle.

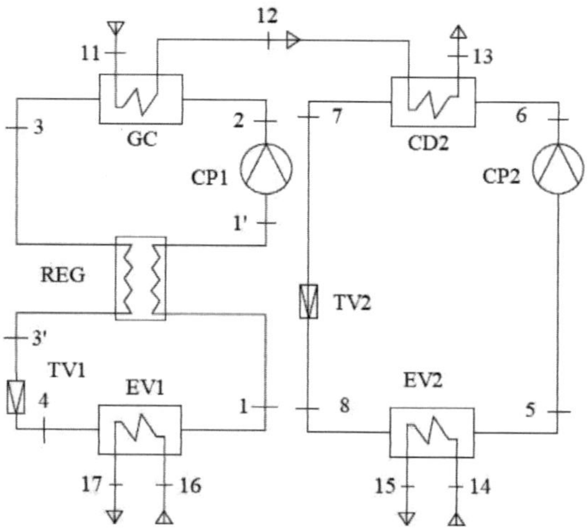

Figure 6. Diagram of the equipment with regenerative subcooling at the refrigeration cycle and subcooling at the heat pump cycle

In this case the exergy balance equation becomes:

$$|\dot{W}_1| + |\dot{W}_2| + \left|\dot{Ex}_{Q_{14-15}}^{T_{14-15}}\right| = \left|\dot{Ex}_{Q_{16-17}}^{T_{16-17}}\right| + \left|\dot{Ex}_{Q_{11-12}}^{T_{11-12}}\right| + \left|\dot{Ex}_{Q_{12-13}}^{T_{12-13}}\right| + \dot{I}_{CP1} + \dot{I}_{\Delta TGC} + \dot{I}_{TV1} + \dot{I}_{\Delta TEV1} + \dot{I}_{CP2} + \dot{I}_{\Delta TCD2} + \dot{I}_{TV2} + \dot{I}_{\Delta TEV2} + \dot{I}_{\Delta TREG} \quad (12)$$

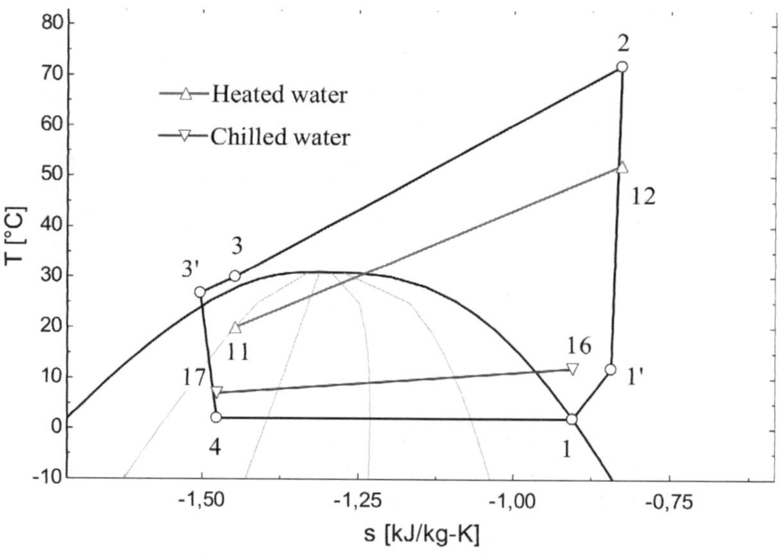

Figure 7. T-s diagram of the refrigeration cycle, cycles connected on the hot water side and regenerative subcooling

When applying the regenerative subcooling in the case of the refrigeration cycle and subcooling at the level of the agent for the heat pump, the working diagrams in t-s coordinates can be followed in figures 7 and 8.

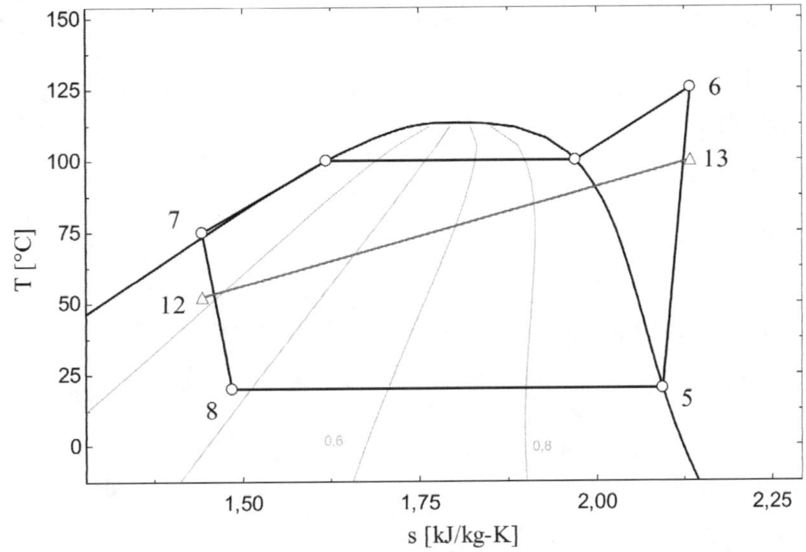

Figure 8. T-s diagram of the heat pump cycle, cycles connected on the hot water side and regenerative subcooling

Compare figure 4 with figure 9 when applying regenerative subcooling at the level of the refrigeration cycle and subcooling at the level of the agent in the case of the heat pump cycle, the exergy damage from the throttling processes is diminished due to the reduction of the temperature difference between the inlet and outlet gate from the throttling valve.

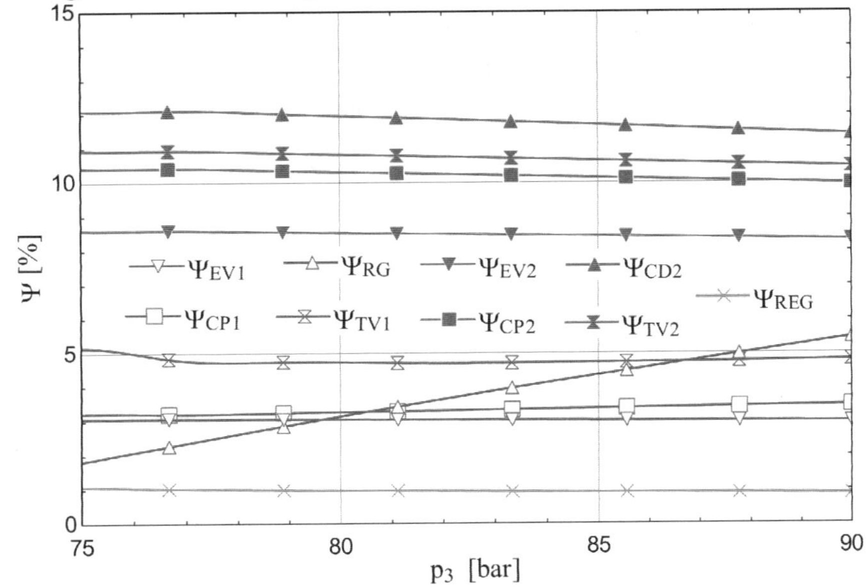

Figure 9. Exergy destruction in each device and process, Cycles connected on the hot water side and regenerative subcooling at the refrigeration cycle and subcooling of the agent at the heat pump cycle

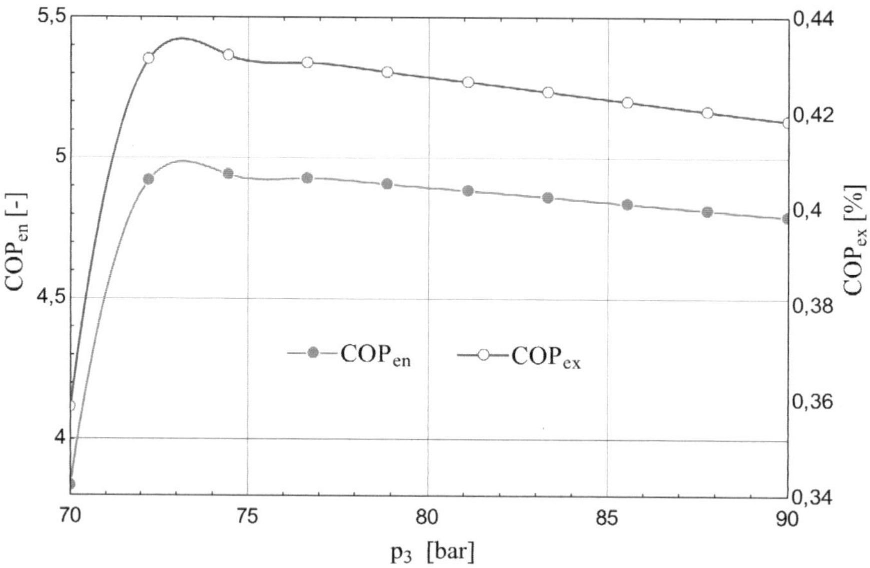

Figure 10. Performance indices COP_{en} and COP_{ex} in the case of cycles connected on the hot water side and regenerative subcooling at the refrigeration cycle and agent subcooling at the heat pump

Following figures 5 and 10 where the performance indices are presented in the case of the two equipments, it is observed the importance and the value of the information offered by the exergetic analysis that led to the significant increase of the performance indices.

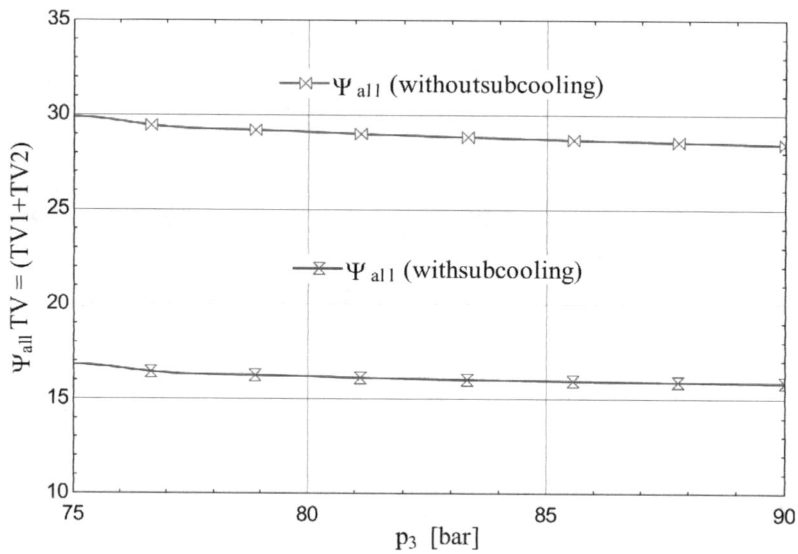

Figure 11. Total representation of exergy destruction in throttling processes for without subcooled cycles and subcooled cycles

According to figure 11 where the summed exergy destruction of the refrigeration cycle and of the heat pump cycle is presented in the case of the two described equipments, with subcooling and without subcooling, a significant reduction of the total exergy destruction from the throttling processes can be

IOP Conf. Series: Materials Science and Engineering 1290 (2023) 012008 doi:10.1088/1757-899X/1290/1/012008

observed. This improvement is possible because the exergy analysis shows the structural and functional dysfunctions and imperfections at the level of each device and process.

Only exergy analysis has the ability to provide information from the inside of the system and offers optimization strategies and priorities.

5. Conclusions

The performances of an equipment that is able to generate hot utility and cold utility were investigated.

The paper focused on identifying the greatest exergy destructions and identifying solutions to reduce them.

The study states that the discharge pressure from the refrigeration cycle with carbon dioxide has a great influence on the performance of the equipment.

The results of the study show that there is an optimal discharge pressure in the case of carbon dioxide.

When applying regenerative subcooling at the refrigeration cycle and subcooling at the heat pump, the exergy destruction in the case of throttling processes is reduced.

Exergy analysis makes it possible to enter the boundaries of the system and highlights the extent of failures associated with each device and process.

Exergy analysis is a powerful thermodynamic optimization tool that indicates priority processes and devices for structural or functional optimization.

Acknowledgements

This work has been funded by the European Social Fund from the Sectorial Operational Programme Human Capital 2014-2020, through the Financial Agreement with the title "Training of PhD students and postdoctoral researchers in order to acquire applied research skills - SMART", Contract no. 13530/16.06.2022 - SMIS code: 153734.

Nomenclature

COP- coefficient of performance [-]

$\dot{Ex}$ - exergy [kW]

h- specific enthalpy [kJ/kg]

$\dot{I}$ - current of exergy destruction [kW]

$\dot{m}$ – mass flow [kg/s]

$\dot{Q}$ - heat flow [kW]

T - thermodynamic temperature [K]

$\dot{W}$ - mechanical power exchange [kW]

Subscripts

CD - condenser

CP – compressor

en – energy

ex – exergy

GC – gas cooler

REG - regenerative heat exchanger

TV - throttling valve

EV - evaporator

Greek symbols

Ψ- exergy destruction factor defined as the ratio between exergy destruction and the total exergy input [%]

IOP Conf. Series: Materials Science and Engineering 1290 (2023) 012008 doi:10.1088/1757-899X/1290/1/012008

References

[1] https://www.consilium.europa.eu/ro/meetings/international-summit/2021/11/01/

[2] Regulation (EU) No 517/2014 of the European Parliament and of the Council of 16 April 2014 on fluorinated greenhouse gases and repealing Regulation (EC) No 842/2006

[3] Saikawa M and Koyama S 2016 Thermodynamic analysis of vapor compression heat pump cycle for tap water heating and development of CO2 heat pump water heater for residential use Appl. Therm. Eng. 106 1236-43

[4] Institut international du froid 2019 8th Conference on Ammonia and CO2 Refrigeration Technology (Paris: IIF-IIR) Online: https://iifiir.org/en/fridoc/6077

[5] Jian L ZhenY Hailong L Shuozhuo H Yuanyuan D and Jinyue Y 2021 Optimal schemes and benefits of recovering waste heat from data center for district heating by CO2 transcritical heat pumps Energy Conversion and Management 245

[6] Xu Y Z Gao J T Bin H and Wang R Z 2022 Multi-criterion comparison of compression and absorption heat pumps for ultra-low grade waste heat recovery Energy 238

[7] Baomin D Xiao L Shengchun L Yingying Z Dan Z Yining F Victor N and Ying H 2020 Dualpressure condensation high temperature heat pump system for waste heat recovery: Energetic and exergetic assessment 218

[8] Wei-wei Y Xinq-qi C Ya-ling H and Fei-yu Y 2017 Theoretical study of a high-temperature heat pump system composed of a CO2 transcritical heat pump cycle and a R152a subcritical heat pump cycle Applied Thermal Engineering 120 pp 228-238

[9] https://www.honeywell-refrigerants.com/europe/product/solstice-zd/

[10] https://www.linde-gas.ro/shop/ro/ro-ig/gaze-imbuteliate/freoni/agent-frigorific-r1234yf-but56kg-p5400106

[11] Radcenco V Criterii de optimizare a proceselor termice Editura Tehnica, Bucuresti, 1977.

[12] Dobrovicescu A Analiza exergetica si termoeconomica a sistemelor frigorifice si criogenice , Editura AGIR, Bucuresti, 2000

[13] Radcenco V Porneala S Dobrovicescu A 1983 Procese in instalatiile frigorifice Editura Didactica si Petagogica, Bucuresti

[14] https://www2.le.ac.uk/ offices / itservices / ithelp / my-computer / programs / ess-10-2 Engineering Equation Solver (verses 10.2 – 2017)

IOP Conf. Series: Materials Science and Engineering 1290 (2023) 012009 doi:10.1088/1757-899X/1290/1/012009

Theoretical aspects of the use of hydrogen in the diesel engine

A Cernat[1], N Negurescu[1], C Pana[1] and C Nutu[2]

[1]Department of Thermotechnics, Engines, Thermal and Frigorific Equipment, University Politehnica of Bucharest, Bucharest, Romania
[2]Department of Automotive, Engines, Thermal and Frigorific Equipment, University Politehnica of Bucharest, Bucharest, Romania

cernatalex@yahoo.com

Abstract. The promotion of fuels with low carbon content is the solution to reduce pollutant emissions in all areas of transport in a relatively short time. In this context, hydrogen is privileged, being a viable alternative fuel for internal combustion engines in all fields of transport, having the advantage of preserving the basic structure of the engine. The main obstacles to the large-scale expansion of the use of hydrogen as a fuel in internal combustion engines in transport are determined by the cost of production, storage on board the vehicle and the distribution network, large investments being required. The concept of using hydrogen as an alternative fuel to the diesel engine is relatively recent, the concerns of specialists being more intense in the last two decades. The improvement of the energy and pollution performance of the diesel engine when operating in a dual system, with diesel and hydrogen, is determined by the improvement of the combustion process due to the good combustion properties of hydrogen. The paper presents theoretical modeling results regarding the influence of hydrogen on the combustion pressure, the pressure rise rate, the heat release rate, the indicated efficiency and the nitrogen oxides emission level.

1. Introduction

The use of different alternative fuels may represent a good solution to reduce the pollutant emissions level of the automotive equipped with diesel engines, fact which may assure the maintenance of these vehicles in service. The promotion of fuels with low carbon content is the solution to reduce pollutant emissions in all areas of transport in a relatively short time. In this context, hydrogen is privileged, being a viable alternative fuel for internal combustion engines in all fields of transport, having the advantage of preserving the basic structure of the engine. The main obstacles to the large-scale expansion of the use of hydrogen as a fuel in internal combustion engines in transport are determined by the cost of production, storage on board the vehicle and the distribution network, large investments being required. The concept of using hydrogen as an alternative fuel to the diesel engine is relatively recent, the concerns of specialists being more intense in the last two decades. The improvement of the energy and pollution performance of the diesel engine when operating in a dual system, with diesel and hydrogen, is determined by the improvement of the combustion process due to the good combustion properties of hydrogen.

Content from this work may be used under the terms of the Creative Commons Attribution 3.0 licence. Any further distribution of this work must maintain attribution to the author(s) and the title of the work, journal citation and DOI.
Published under licence by IOP Publishing Ltd

IOP Conf. Series: Materials Science and Engineering 1290 (2023) 012009 doi:10.1088/1757-899X/1290/1/012009

Because hydrogen has very wide inflammability limits and high combustion speed, the homogeneous air-hydrogen mixture burns completely even at very low dosages, improving the engine's pollution and energy performance.

When using hydrogen, emissions of greenhouse gases and pollutants are reduced, influences dependent on the proportion of hydrogen used to replace diesel.

When increasing the percentage of hydrogen to replace a larger part of the diesel dose, the air-hydrogen mixture becomes richer and abnormal combustion manifestations, spontaneous ignition from hot sources in the combustion chamber, violent combustion and detonation can be favored. These manifestations are favored by the fact that hydrogen has a low minimum ignition energy, a short extinguishing distance to the wall and a very wide flammability range. These aspects can constitute limits for establishing the percentage of substitution of diesel fuel with hydrogen.

Different researches study the influence of hydrogen use as fuel for diesel engines. As example, Santoso [1] develops a study based on AVL modelling software in order to evaluate better the combustion process into a diesel engine fuelled with diesel fuel and hydrogen. Santoso [1] uses different hydrogen flows to fuel the diesel engine reaching hydrogen ratio of 0.2, 0.3 and 0.5. Indicated efficiency of the engine has a superior value for low hydrogen amounts, but for large hydrogen flows the indicated efficiency starts to decrease slowly [1]. Higher hydrogen flows use lead to the reduction of the maximum pressure and pressure rise rate, Santoso [1] explains that the reduction of the diesel fuel being the main issue of this aspect. Heat release rate starts to increase for low hydrogen quantities the maximum value being achieved sooner per cycle, but higher hydrogen amounts leads to the reduction of the heat released during combustion. More hydrogen leads to the increase of the combustion duration. For lower hydrogen flows, the appropriation between experiment and modelling results is better comparative to situation of high hydrogen flows. The rate of heat release appears in higher values especially for low hydrogen quantities and low amounts of hydrogen assure the acceleration of the combustion and the increase of the heat amount released per cycle comparative to high amounts of hydrogen. Santoso [1] shows that higher amounts of hydrogen leads to the increase of combustion temperature especially at the end of combustion and lower hydrogen amount assure the increase of the temperature at the beginning of the combustion process. Koten [2] explains that hydrogen use leads to the increase of the thermal efficiency, but NOx emission level increase at high loads. The soot emission level is decreased at hydrogen use, the decrease being more significant for large hydrogen amounts [2]. Liu [3] observes that hydrogen use assures the increase of the thermal efficiency of a single cylinder engine with 13% but leads to the increase of NOx. Zhang [4] uses natural gas and carbon dioxide to fuel a single diesel engine, hydrogen being used as addition fuel. Zhang [4] and shows that the increase of hydrogen percent in the gaseous fuel leads to the increase of the indicated thermal efficiency for all types of fuel, especially for the fuels with low natural gas content. Looking on combustion Zhang observes the increase of the pressure peak increase once with the rise of the hydrogen cyclic quantity, from 4 to 5 MPa. The maximum pressure rate the tendency is similar, the peak values being higher for increased values of hydrogen, the rite rise increasing from 0.09 MPa/CAD till 0.15 MPa/CAD. In generally the NOx emission level starts to increase at the rise of the hydrogen addition, but for the fuels that admit more hydrogen in their composition the NOx emission level in lower comparative to the fuels which has lower hydrogen content [4]. Lean mixtures operation, λ=1.4…1.7, is much better to use in terms of NOx reduction if low hydrogen quantities are used [4]. Demirci [5] uses small amounts of hydrogen to fuel a Lombardini diesel engine and observe the increasing of the thermal efficiency due to improved combustion at hydrogen use. The NOx emission level increase with the increase of engine load, but for each operating regime, the increase of the hydrogen amount the NOx emission level decreases for all small loads [5]. In general, at hydrogen use the NOx emission levels were decreased for hydrogen use comparative to diesel fuel use [5]. Increasing of the hydrogen ammount leads also to the reduction of the smoke emission level [5]. Dabnath [6] investigates the use of hydrogen as gaseous fuel for a diesel engine and shows the increase of the brake thermal efficiency at the increase of the hydrogen percent,

IOP Conf. Series: Materials Science and Engineering 1290 (2023) 012009 doi:10.1088/1757-899X/1290/1/012009

for all engine loads, but volumetric efficiency decreases [6]. Thanigaivelan [7] try to ameliorate the performance of the diesel engine fueled with biodiesel by using hydrogen as supplementary fuel. Addition of hydrogen leads to the increase of the thermal efficiency comparative to diesel fuel use and to biodiesel fuel operation [7]. For low loads the NOx emission level decreases for hydrogen addition comparative to low hydrogen quantities [7]. Bari [8] shows the effect of hydrogen on diesel engine performance, highlighting the increase of the thermal efficiency at the increase of the hydrogen cyclic dose, but the NOx emission level is increased for large hydrogen doses. The fuel saving percent starts to increase significantly once with the rise of the hydrogen percent [8]. Sughayyer [9] shows that the use of hydrogen leads to significant reduction of diesel fuel and reduction of the emission levels in terms of engine efficiency maintaining, developing a modelling procedure in order to make a correlation with the experimental results. Sabeghi [10] shows the reduction of the NOx emission level with 11% at the use of HHO as fuel comparative to diesel fuel operation. Raju [11] shows that hydrogen injection leads to the reduction of the smoke emission level but not for the NOx emission level. The brake thermal efficiency, in-cylinder pressure and the heat release rate starts to increase once with the rise of hydrogen cyclic dose [11]. Castro [12] observes the reduction of the NOx emission level at the increase of the hydrogen quantity especially at low engine loads, but the smoke emission level was decreased for all operating regimes by hydrogen addition. Castro [12] shows that hydrogen can assure he reduction of the diesel fuel consumption without combustion un-normal phenomena's. Akar [13] explains the influence of hydrogen addition on improving performance of a waste oil fueled diesel engine, showing the increase of the brake thermal efficiency, reduction of the carbon emissions levels at hydrogen use, but increasing of the NOx emission level in correlation with the increasing of the combustion temperature. Zhang [14] observes the increase of the brake thermal efficiency at hydrogen use as additional fuel to a diesel engine fueled with biodiesel, up till 38%, and a slightly reduction of the NOx emission level. Wang [15] limits the hydrogen quantity in order to limit the increase of the NOx emission level at a diesel engine, even the engine performance are improved by hydrogen use and try to explain the apply of HCCI method for pollutant emission improve based on literature review. Thiyagarajans [16] explains the effects of the hydrogen at fuelling a diesel engine in order to reduce the carbon emissions, reduction of the smoke emission, showing that improved combustion leads to the NOx emission level increase and suggests EGR or catalytic converter use to decrease the NOx emission level. Karimi [17] shows the improvement of the diesel engine performance at the use of hydrogen and oxygen in terms of engine efficiency, with 1.6%, reduction of the soot with 72% and the increase of NOx emission level with 63%. Further, Karimi [17] uses EGR method to decrease the increased NOx emission level, with 79% if oxygen is used too, observing also a slightly increase in engine performance, with 2.6% for brake thermal efficiency. Rorimpandey [18] shows the peak of the heat release rate begins to rise during the combustion of the hydrogen comparative to the diesel fuel and n-heptane. Liu [19] develops a study based on modelling showing that pre-chamber hydrogen combustion may leads to the increase of the thermal efficiency and reduction of the NOx emission level for a diesel engine, even the pressure rise rate during combustion is increased by hydrogen use. Duarte [20] shows the reduction of the NOx emission level at low loads for a diesel engine fueled with hydrogen. But the NOx emission level increases at medium and high engine loads [20]. The particulate matter is reduced also by hydrogen use for all engine loads, once with the increase of hydrogen amount the particulate matter level being decreased with 12% [20].

The paper presents theoretical modeling results regarding the influence of hydrogen on the combustion pressure, the pressure rise rate, the heat release rate, the indicated efficiency and the nitrogen oxides emission level.

2. Thermodynamic modelling of the combustion process

For the study of the processes in the cylinder of the diesel engine fueled with conventional fuel, diesel fuel, and in the dual system with diesel fuel and hydrogen, a one-zone, zero-dimensional type thermodynamic model was developed. The state of the engine fluid in the engine cylinder is determined based on energy and mass conservation equations, specific to each process, intake, compression, expansion after combustion and exhaust. Triggering the calculation of the gas exchange requires knowledge of the state parameters of the gas in the cylinder at the moment of opening the exhaust valve - α_{OEV}. These initial parameters are chosen (values determined by the approximate cycle calculation method can be used) and after completing the cycle, through iterations, their final values are established [21]. The general relation of cylinder volume is used also for moment of opening the exhaust valve are:

$$V = V_S \cdot \left\{\frac{1}{(\varepsilon-1)} + 0.5 \cdot [(1 - cos\alpha) + \Lambda/4 \cdot (1 - cos2\alpha)]\right\} \quad (1)$$

where: V_s is cylinder volume, ε-compression ratio.
The substance quantity per cycle is:

$$N_c = \frac{p \cdot V}{R_M \cdot T} \quad (2)$$

The in-cylinder temperature variation during inlet-exhaust processes is evaluated with the formula:

$$\frac{dT}{d\alpha} = \frac{1}{N_c \cdot C_{vc}} \cdot \left\{\left(i_A|_{T_A} - i_C|_T\right) \cdot \frac{dN_A}{d\alpha} + \left(i_E|_{T_E} - u_C|_T\right) \cdot \frac{dN_E'}{d\alpha} - (i_C|_T - u_C|_T) \cdot \left(\frac{dN_E}{d\alpha} + \frac{dN_A'}{d\alpha}\right) - K_{conv} \cdot \left[A_{wall} \cdot (T - T_{wall}) + A_{CH} \cdot (T - T_{CH}) + A_{cyl} \cdot \left(T - T_{cyl}\right)\right] \cdot \frac{1}{6 \cdot n} - p \cdot V_s/2 \cdot (sin\alpha + \Lambda/2 \cdot \sin 2\alpha) \cdot \pi/180\right\} \quad (3)$$

where: N_c, N_A, N_E-in-cylinder, inlet and exhaust substance quantities, N_A', N_E'- inlet and exhaust reverse flows substance quantities, *i*-specific enthalpy, *u*-specific internal energy, c_{vT}-specific heat at constant temperature, K_{conv} - coefficient of convection heat transfer, indices of the area A of heat transfer which are similar for temperatures T: A-admission, c-cylinder, E-exhaust, wall-cylinder wall, CH-cylinder head, n-engine speed, p-in-cylinder pressure, Λ-cinematic parameter. By integrating the equation, the temperature at the end of the calculation interval is obtained. From the general equation of state applied at the end of the calculation interval, the pressure results:

$$p = \frac{N_c \cdot R_M \cdot T}{V} \quad (4)$$

The calculation of the compression process starts when the intake valve is closed. The amount of engine fluid in the cylinder and its composition do not vary during the compression process (dN_A=0; dN_E=0; dN_A'=0; $dN_{E'}$=0 in the differential relation of the temperature (3)).
The combustion process simulation starts with the definition of the fuel dosage per cycle (kg/cycle):

$$\overline{m_f} = \frac{\rho_0 \cdot V_S \cdot \eta_v}{\lambda \cdot L_t} \quad (5)$$

where ρ_o-fresh charge density, volumetric efficiency, λ-air excess coefficient, L_t-theoretically air minimum quantity for stoichiometric combustion.
The energy balance equation during combustion is:

$$\frac{dU}{d\alpha} = \frac{dQ_{comb}}{d\alpha} - \frac{dL}{d\alpha} - \frac{dQ_{wall}}{d\alpha} \quad (6)$$

where: dU/dα is the internal energy variation, $dQ_{comb}/d\alpha$ is the amount of heat released by combustion, dL/dα is the mechanical work, $dQ_{wall}/d\alpha$ is the heat given to the walls on the elementary calculation interval dα.

Each term of the relationship of the energy balance equation during combustion is explained:
Internal energy variation is:

$$\frac{dU}{d\alpha} = n_M \cdot c_v - \frac{dT}{d\alpha} \tag{7}$$

where n_M is the number of kilo-moles of charge in the engine cylinder evaluated based on the molecular mass of the fuel M_f and fuel dose per cycle m_f:

$$n_M = (\lambda \cdot L_t + \frac{1}{M_f}) \cdot (\mu_{ch} + \gamma) \cdot m_f \tag{8}$$

The cyclic fuel dose m_f determined in (kg fuel/cycle) is:

$$m_f = \frac{\rho_0 \cdot V_s \cdot \eta_v}{\lambda \cdot L_t \cdot M_a} \tag{9}$$

the density of the fresh charge ρ_0 being calculated from the equation of state under the conditions (p_0, T_0) in the case of the engine with normal intake.
Considering the two combustion mechanisms – combustion of preformed mixtures (rapid combustion) and diffusive combustion (moderate combustion) – the combustion characteristic of the diesel engine can be expressed by combining two Vibe-type relations:

$$\xi = R_c \times \left[1 - e^{-6,9 \times x_r^{m_p^*+1}}\right] + (1 - R_c) \times \left[1 - e^{-6,9 \times x_d^{m_d^*+1}}\right] \tag{10}$$

where R_c expresses the fraction of the available heat released in the rapid combustion phase, m_p* and m_d* - kinetic exponents, x_r and x_d - relative angles determined based on the moments of start and durations of the rapid and diffusive combustion, [21].

Mechanical work performed is:

$$\frac{dL}{d\alpha} = p \cdot \frac{dV}{d\alpha} \tag{11}$$

The heat transferred to the walls:

$$\frac{dQ_{wall}}{d\alpha} = K_{conv} \cdot A_{wall} \cdot (T - \overline{T_{wall}}) \cdot \frac{1}{6 \cdot n} \tag{12}$$

The amount of heat released by combustion:

$$\frac{dQ_{comb}}{d\alpha} = m_f \cdot Q \cdot \frac{d\xi}{d\alpha} \tag{13}$$

Q being the heat of reaction of the fuel (in the case of the diesel engine where $\lambda>1$, Q represents the lower heating value of the fuel). The combustion characteristic $d\xi/d\alpha$ is approximated by a double Vibe function:

$$\frac{d\xi}{d\alpha} = R_c \cdot f_1 + (1 - R_c) \cdot f_2 \tag{14}$$

where f_1 and f_2 are combustion parameters for rapid and diffusive combustion phases, respectively [21].

Using the relationship of the energy balance equation, is obtained:

$$\frac{dT}{d\alpha} = \frac{k-1}{n_M \cdot R_M} \cdot \left(\frac{dQ_{comb}}{d\alpha} - \frac{Q_{wall}}{d\alpha} - \frac{dL}{d\alpha}\right) \tag{15}$$

Taking the calculation interval $\Delta\alpha$=1...3 CAD, the temperature of the current step is determined based on the knowledge of the previous one:

$$T_2 = T_1 + \left(\frac{dT}{d\alpha}\right) \cdot \Delta\alpha \tag{16}$$

The pressure of the respective step results from the equation of state:

$$p = \frac{n_M \cdot R_M \cdot T}{V} \tag{17}$$

The expansion process is evaluated from the energy balance equation on the expansion evolution, is obtain:

$$\frac{dT}{d\alpha} = \frac{1}{N_C \cdot C_V} \cdot \left(\frac{Q_{wall}}{d\alpha} + \frac{dL}{d\alpha}\right) \tag{18}$$

The amount of gas in the cylinder:

$$N_C = (N_g + N_f) \cdot \overline{m_f} \tag{19}$$

The most important diesel engine emissions are emissions of NOx and smoke. The estimation of those two emission level is made with the mathematical model. Emission level of the NO is estimated with the relation proposed by Schroer [21]:

$$NO\ [ppm] = C_1 \cdot (\lambda - 0.89)^{2.11} \cdot n^{-0.39} \cdot Q^{2.49} \cdot m_a^{-5.01} \cdot \left(\int_0^Q T^{11} \cdot p^9 \cdot dQ\right)^{0.28} \tag{20}$$

where: λ- coefficient of air excess [-], n-speed [rev/min], Q-quantity of heat release [kJ/cycle], m_a-fresh charge mass [g], T-temperature [K], p-pressure [bar], C_1- constant, $C_1 = 0.24 \cdot 10^{-10}$, $[(rot/s)^{0.39} \cdot g^{5.01} \cdot ppm / kJ^{2.77} \cdot K^{3.08} \cdot bar^{2.52}]$

The smoke mass formation speed is evaluated with the formula proposed by Hiroyasu [22]:

$$m_{sf}\left[\frac{mg}{s}\right] = 2.5 \cdot 10^{-3} \cdot m_{tf} \cdot p^{0.5} \cdot \exp\left(-\frac{52298}{R \cdot T}\right) \tag{21}$$

where: m_{tf} is total fuel mass, p and T are pressure and temperature during combustion

Oxidation of the smoke is estimated by the speed relation proposed by Morel-Karibel [23]:

$$m_{so}\left[\frac{mg}{s}\right] = 9.55 \cdot 10^{-4} \cdot m_{soot} \cdot p^{0.5} \cdot \exp\left(-\frac{5000}{T}\right) \tag{22}$$

In the exhaust gases the smoke emission level is determined as difference between formation and oxidation laws [22, 23].

The thermo-gas-dynamic processes in the cylinder are simulated for an experimental single-cylinder compression ignition engine with the following characteristics: cycle τ=4 strokes, bore D=85 mm, stroke S=115 mm, compression ratio ε=7-23, total displacement V_t=0.652 dm^3, operating speed n=900 rpm, advance when producing diesel injection $\beta_{injection}$=13 CAD, standard diesel fuel consumption 13 cm^3 per minute, compression ratio 13, liquid cooling.

Considering that the operation with a reduced addition of hydrogen, limited to about 10%, is considered, the engine is supplied with hydrogen in the intake system, the hydrogen being introduced into the intake air.

The energetic substitute ratio, XE, is established based on cyclic amount of used fuels, diesel fuel and hydrogen, and their lower heating value.

3. Results

The following figures present the results of modelling like variation of the in-cylinder pressure during cycle for different substitute ratios, maximum pressure rise rate, heat release rate during combustion, thermal efficiency and level of the NO emission.

In the figure 1 the pressure variations versus crank angle are shown.

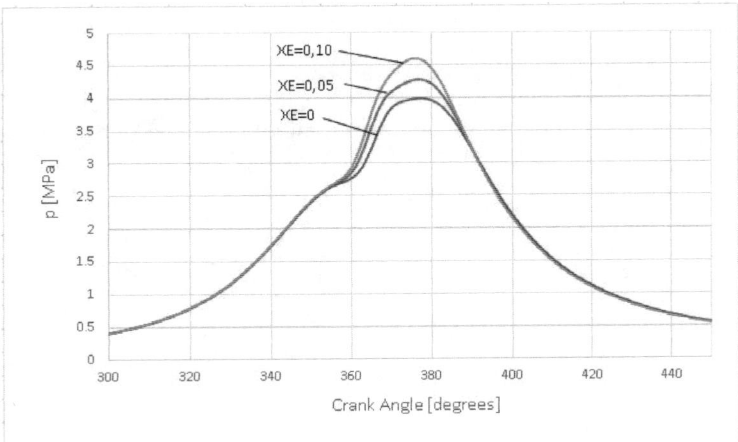

Figure 1. Pressure diagrams at fuelling with diesel fuel and diesel fuel and hydrogen.

At hydrogen use the in-cylinder pressure starts to increase, figure 1, and is achieved sooner per cycle, with 1 crank angle degree for XE=0.10, figure 3, due to reduction of the combustion duration. The maximum pressure increases with 7.4% at XE=0.05 and with 15.3% at XE=0.10, as figure 2 shows.

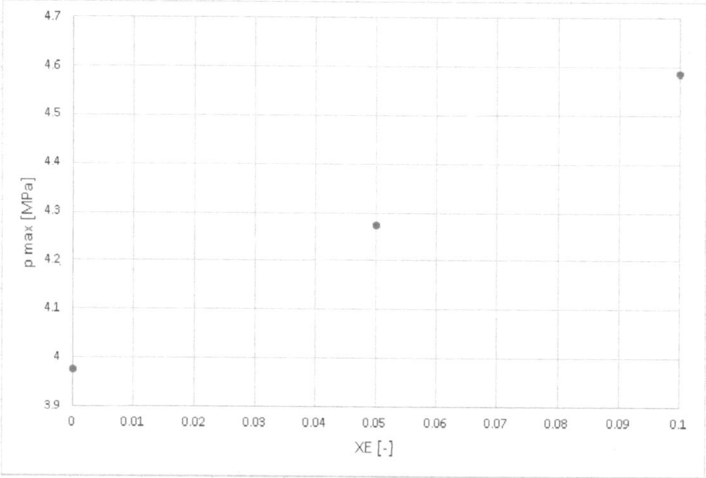

Figure 2. Maximum pressure at fuelling with diesel fuel and diesel fuel and hydrogen

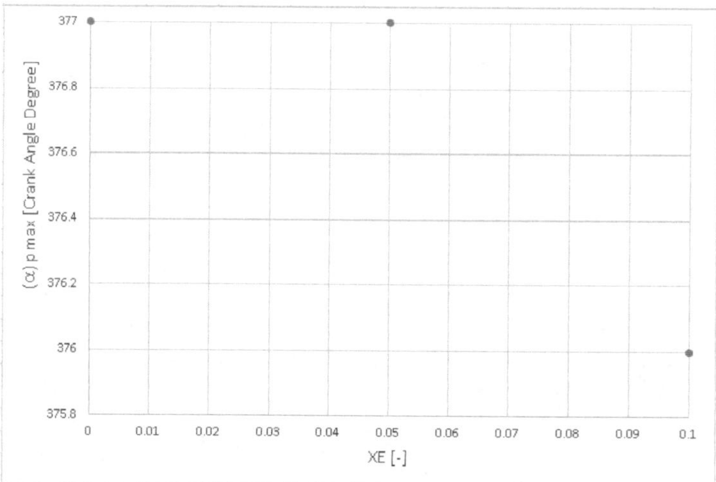

Figure 3. Angle of maximum pressure at fuelling with diesel fuel and diesel fuel and hydrogen.

The maximum pressure rise rate, figure 4, starts to increase at hydrogen use, with 15.3% at XE=0.05 from 0.147 MPa /CAD to 0.170 MPa/CAD. For XE=0.10, the increase is around 28%, from 0.147 MPa/CAD to 0.189 MPa/CAD. The values do not exceed the usual area which assure the normal engine operation.

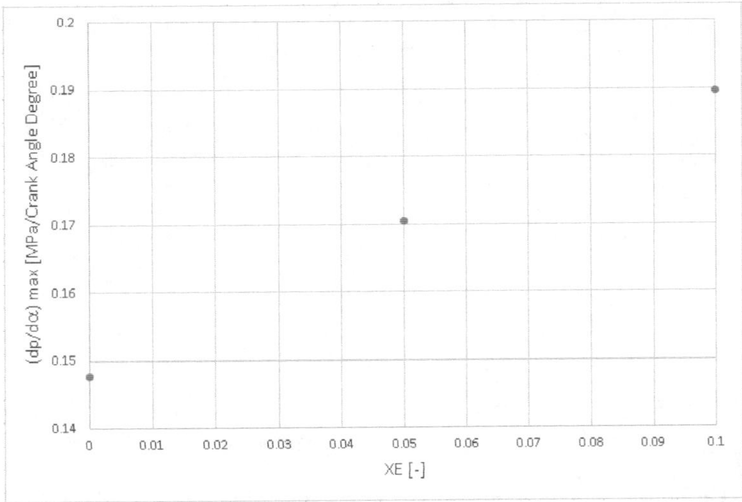

Figure 4. Maximum Pressure Rise Rate at fuelling with diesel fuel and diesel fuel and hydrogen.

The heat release rate during combustion of air-diesel fuel-hydrogen mixtures is superior to the value registered for classic fuelling, figure 5. At XE=0.05 the peak value of the heat release rate is with 2.3% higher comparative to classic fuelling and for XE=0.10 the increase is around 9.3% comparative to classic fuelling. The angle of maximum heat release rate tends to appears sooner per cycle in correlation with the higher combustion speed and lower heating value of hydrogen versus diesel fuel, figure 5.

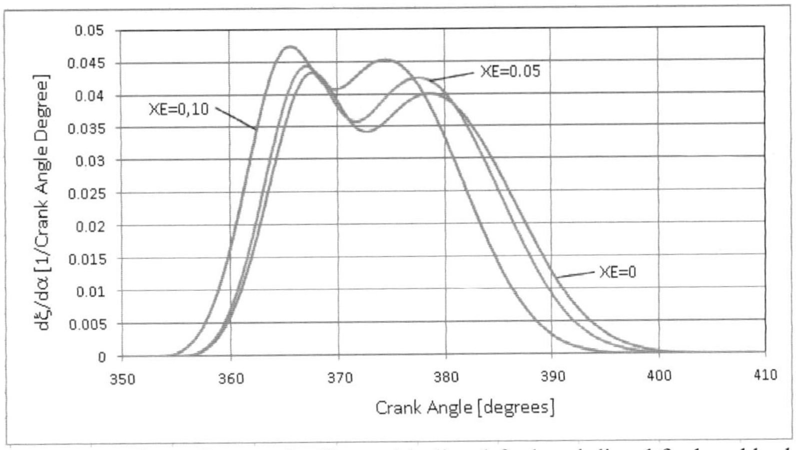

Figure 5. Heat Release Rate at fuelling with diesel fuel and diesel fuel and hydrogen.

Higher value of the lower heating value of hydrogen influences the combustion temperature, which starts to increase at hydrogen use. At XE=0.05 the peak value of the combustion temperature increases with 1%, due to that fact that lower amount of hydrogen is used, but for XE=0.10, the peak temperature is with 2.5% higher comparative to diesel fuel operation, figure 6 and 7. At hydrogen use, the angle at which maximum pressure occurs appears sooner per cycle, with 3 CAD closer to TDC for XE=0.05 and with 5 CAD closer to TDC for XE=0.10, figure 6.

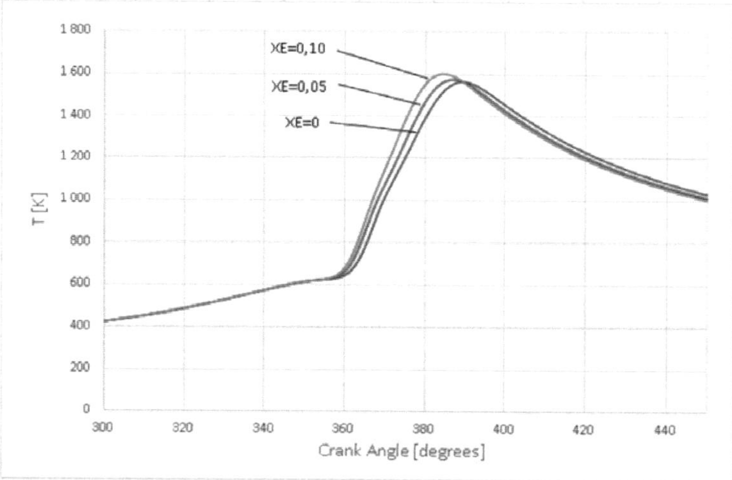

Figure 6. Combustion temperature at fuelling with diesel fuel and diesel fuel and hydrogen.

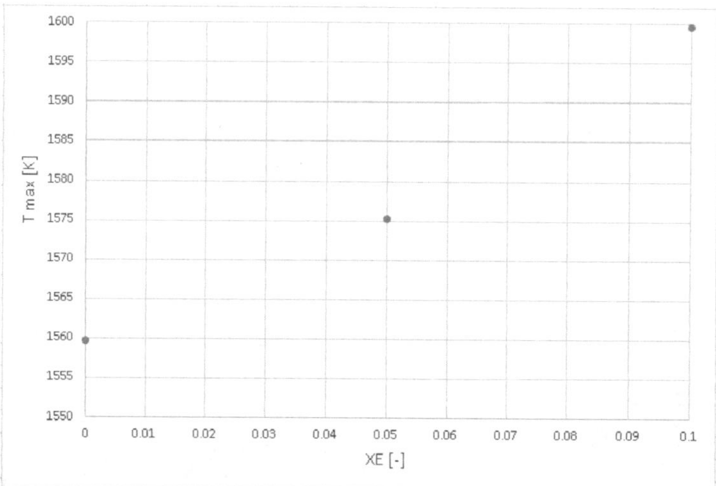

Figure 7. Maximum combustion temperature at fuelling with diesel fuel and diesel fuel and hydrogen.

Good combustion properties of hydrogen, like higher combustion speed, higher lower heating value, larger in-flammability limits leads to the increase of the indicated thermal efficiency of the engine at dual fuelling, figure 8.

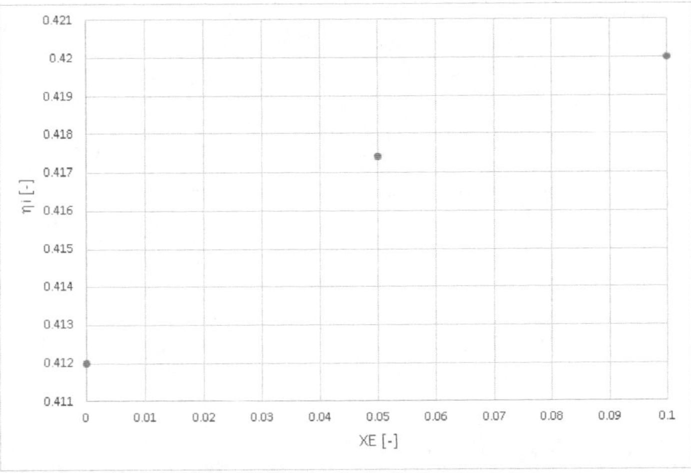

Figure 8. Indicated Thermal Efficiency at fuelling with diesel fuel and diesel fuel and hydrogen.

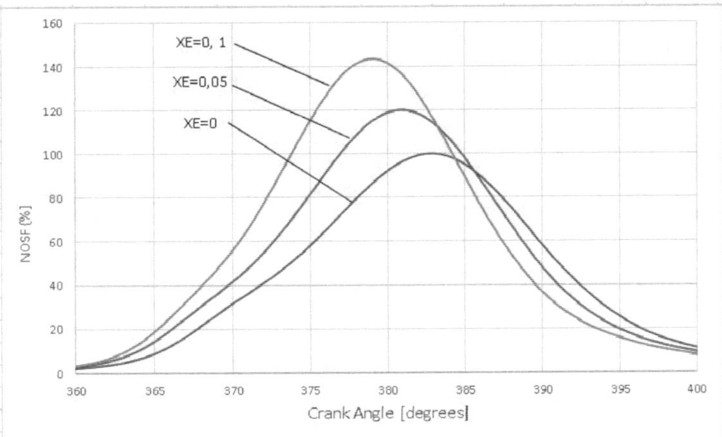

Figure 9. The NO speed forming for different energetic substitute ratios.

The presence of hydrogen in the engine inlet leads to the reduction of the available oxygen which influence the reduction of the nitrous oxides formation quantity. At hydrogen use, the value of the λ decrease from 1.98 till 1.88 at XE=0.05 and 1.79 at XE=0.10, showing the reduction of the available in-cylinder oxygen. The speed of NO formation is increased in the presence of hydrogen, with 19% at XE=0.05 and with 43% at XE=0.10, figure 9, but the final NO quantity is reduces comparative to classic fuelling figure 10. The reduction of the diesel fuel quantity reduce the amount of diesel fuel which burns during premixed phase of combustion and the final NO quantity decreases with 5.2% at XE=0.05 and with 12.8% at XE=0.10. Also, the acceleration of the combustion process at hydrogen use may leads to the reduction of the NO emission level.

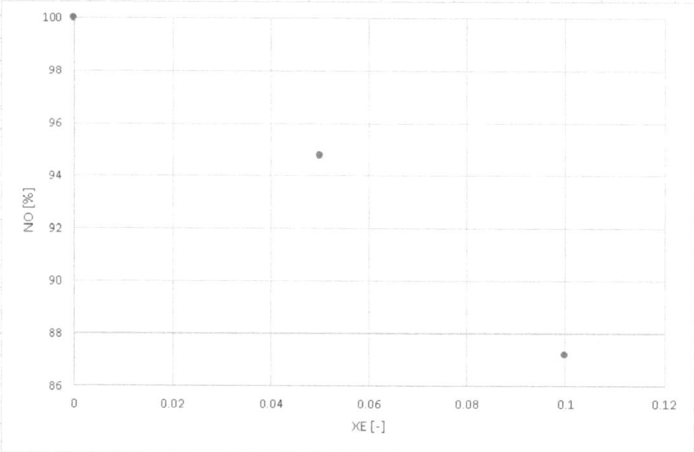

Figure 10. The NO emission level estimation for different energetic substitute ratios.

The maximum of the soot forming speed decreases with 5.2% at XE=0.05 and with 8.3% at XE=0.10 due to reduction of the carbon content in the in-cylinder mixture at the increase of the hydrogen cyclic dose, figure 11.

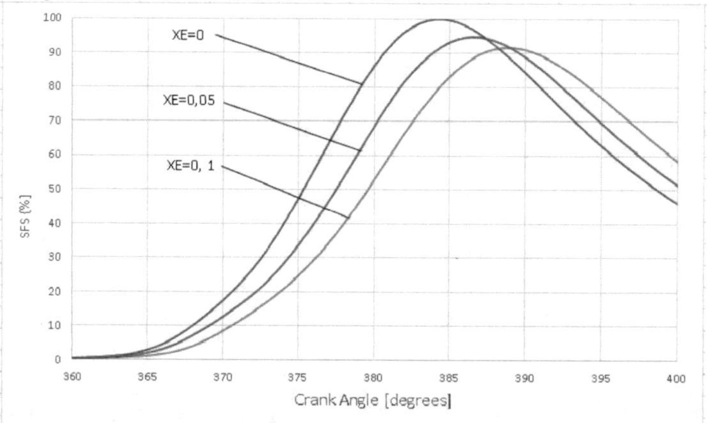

Figure 11. Smoke forming speed for different energetic substitute ratios.

At hydrogen use, the maximum of the soot oxidation speed increases with 7.5% at XE=0.05 and with 19.7% at XE=0.10 comparative to classic fuelling, XE=0, due to acceleration of the combustion process under the influence of higher combustion speed and lower heating value of hydrogen, figure 11.

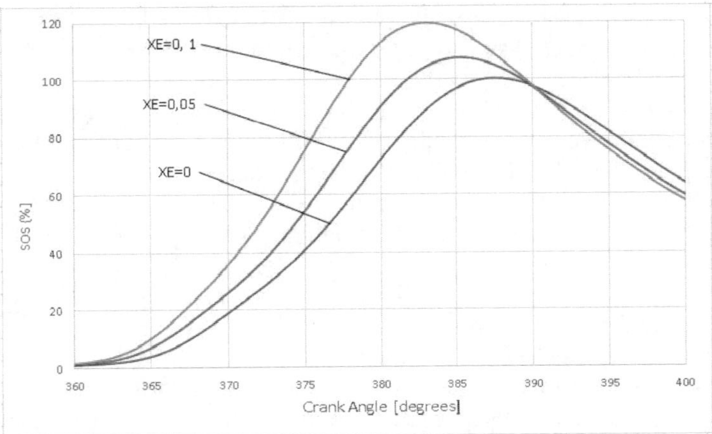

Figure 12. Smoke oxidation speed for different energetic substitute ratios.

In conditions of lower soot formation and acceleration of the soot oxidation at hydrogen use, the estimated smoke emission starts to decrease at hydrogen use, with 0.78% at XE=0.05 and with 1.3% at XE=0.10 comparative to diesel fuel use, figure 13.

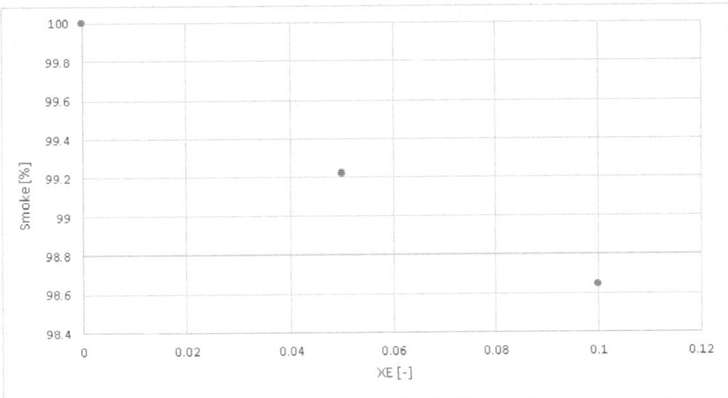

Figure 13. Smoke emission level estimation for different energetic substitute ratios.

4. Conclusions

By applying the physical-mathematical model developed for modeling the thermo-gas-dynamic processes in the engine cylinder, the effect of the use of hydrogen in diesel engines was analyzed on the main quantities that characterize the evolution of the engine fluid during the cycle (variation of pressure and temperature) and the combustion characteristics on which performance depends energy and pollution of the engine. The following conclusions are obvious:

• the maximum pressure increases with the increase in the percentage of hydrogen addition in the inlet air: if with the increase in hydrogen consumption, the dose of diesel fuel was reduced accordingly, keeping unchanged the amount of heat released during combustion process, the maximum pressure increases from 3.98 MPa at fuelling with diesel fuel, to 4.58 MPa when the addition of hydrogen in the inlet air is XE=0,10, as a result of the reduction of the combustion duration;

• the autoignition delay is reduced to the increase in the percentage of hydrogen addition due to the higher flammability of hydrogen compared to diesel fuel, which causes the maximum pressure peak to be closer to the TDC, under the conditions that the diesel fuel injection timing unchanged;

Due to the high combustion speed of hydrogen, the homogeneous air-hydrogen mixtures inside the cylinder burn faster;

• the maximum rate of heat release increases with the increase in the percentage of hydrogen added in the inlet air because hydrogen has an order of magnitude higher combustion rate than other hydrocarbon-based fuels, and in the presence of hydrogen in the mixture the combustion rate of hydrocarbons increases which favors the combustion process. When the amount of heat developed in the combustion process is kept unchanged, the maximum combustion speed increases from 0.043 when operating on diesel fuel, XE=0.0 to 0.047 when the addition of hydrogen is XE=0.10;
• the maximum rate of pressure increase increases with the increase in the percentage of added hydrogen because hydrogen has a higher combustion rate by an order of magnitude compared to other hydrocarbon-based fuels, and in the presence of hydrogen in the mixture the combustion rate of hydrocarbons increases which favors the combustion process: if with the increase in hydrogen consumption, the dose of diesel fuel was reduced accordingly, keeping the amount of heat developed in the combustion process unchanged, the maximum of pressure rise rate increases from 0.147 MPa/CAD when operating with diesel fuel to 0.189 MPa/CAD when the addition of hydrogen is XE=0.10, as a result of the reduction of the combustion duration and the auto-ignition delay;
• the indicated efficiency of the engine increased with the addition of hydrogen in the intake due to the decrease in the duration of combustion and the intensification of combustion because hydrogen has very wide inflammability limits and high combustion speed. When the amount of heat developed in the combustion process is kept unchanged, the indicated efficiency increases by 2.2%, from 0.4109 when operating with diesel fuel, XE=0.0 to 0.42 when the addition of hydrogen is XE=0.10;
• when operating in dual mode with diesel fuel and hydrogen addition in the inlet air, due to the reduction of the available oxygen, reduction of l and reduction of the amount of diesel fuel that burns during premixed phase of combustion, a decrease in NO emission level around 12.8% at XE=0.10 is assured;
• when the engine operates in a dual mode with diesel fuel and hydrogen, due to the lower carbon content and due to complete combustion, the soot formation process is reduced and the soot oxidation process is significantly favored compared to fueling only with diesel fuel. The soot oxidation process is favored when the proportion of hydrogen substitutes diesel fuel increases, which explains the slower continuous decrease in the calculated smoke emission around 1.3% at XE=0.10.

References

[1] Santoso W B, Bakar R A and Nur A 2013 Combustion characteristics of diesel-hydrogen dual fuel engine at low load *Energy Procedia* **32** 3-10 pp3-10
[2] Koten H 2018 Hydrogen Effects on the Diesel Engine performance and emissions International Journal of Hydrogen Energy **43** (22) DOI 10.1016/j.jihydene.2018.04.146
[3] Liu X, Seberry G, Kook S, Chan Q N, Hawkes E 2022 Direct injection of hydrogen main fuel and diesel pilot fuel in a retrofitted single-cylinder compression ignition engine *International Journal of Hydrogen Energy* **47** (84) pp 35864-35876
[4] Zhang X, Hou X S 2019 Experimental research on low calorific value gas blended with hydrogen engine *Energy Procedia* **158** pp 459-464 DOI 10.1016/j.egypro.2019.01.133
[5] Demirci A, Koten H and Gumus M 2018 The effects of the small amounts of hydrogen addition on performance and emissions of a direct injection compression ignition engine *Thermal Science* **22** (00) pp 4-4 DOI 10.2298/TSCI170802004D
[6] Debnath B, Saha U and Sahoo N 2012 Effect of hydrogen-diesel quantity variation on brake thermal efficiency of a dual fuelled diesel engine *Journal of Power Technologies* **92** (1) pp 55-67
[7] Thanigaivelan V, Loganathan M 2019 Performance and emission characteristics of a dual fuel DI engine using cashew nut shell biodiesel and hydrogen as fuel *International Journal of Mechanical Engineering and Technology* **10** (01) (IJMET 10_01_088) pp 847-857
[8] Bari S, Esmaeil M M 2010 Effect of H2/O2 addition in increasing the thermal efficiency of a diesel engine *Fuel* **89** pp 378-383
[9] Sughayyer M 2016 Effects of the Hydrogen Addition on Power and Emissions Outputs from

Diesel Engines *Journal of Power and Energy Engineering* **04** (01) pp 47-56 DOI 10.4236/jpee.2016.41003

[10] Sabeghi M, Moghiman M and Gandomzadeh D 2023 Experimental study of the effect of HHO gas injection on pollutants produced by a diesel engine at idle speed *International Journal of Hydrogen Energy* **48** (24) pp 9117- 9126 DOI.ORD10.1016/j.ijhydene.2022.12.010

[11] Raju P, Masimalai S K, Ganesan N and Karthic S V 2020 Engine's Behavior an hydrogen addition of waste cooking oil fueled light duty diesel engine –A dual fuel approach *Energy* **194** (116844)

[12] Castro N, Toledo M and Amador G 2019 An experimental investigation of the performance and emissions of a hydrogen-diesel dual fuel compression ignition internal combustion engine *Applied Thermal Engineering* **156** pp 660-667

[13] Akar M A, Kekilli E, Bas O et.al.2018 Hydrogen enriched wasteoil biodiesel usage in compression ignition engine *International Journal of Hydrogen Energy* **43** (38) pp 18046-18052

[14] Zhang X, Xia C, AlShalhi M et.al.Assessments of the power productions, energy consumption and emission comparison of hydrogen feed vehicles *Fuel* **334** (Part 2) (126794)

[15] Wang S, Zhang Z, Hou X et.al. 2023 The environmental potential of hydrogen addition as complementation for diesel and biodiesel: A comprehensive review and perspectives *Fuel* **342** (127794)

[16] Thiyagarajan S, Varuvel E, Karthickeyan V et.al. 2022 Effect of hydrogen on compression-ignition (CI) engine fueled with vegetable oils/biodiesel from various feedstocks: A review *International Journal of Hydrogen Energy* **47** (88) pp 37648-37667

[17] Karimi M, Wang X et. Al. Numerical investigation on hydrogen-dieseldual-fuel engine improvements by oxygen enrichment International Journal of Hydrogen Energy 47 (60) pp 25418-25432

[18] Rorimpandey P, Yip H L et.al. Hydrogen-diesel dual-fuel direct injection (H2DDI) combustion under compression-ignition engine conditions *International Journal of Hydrogen Energy* **48** (2) pp 766-783

[19] Liu X, Aljabri H, Silva M et.al.2023 Hydrogen pre-chaber combustion at lean-burn conditions on a heavy-duty diesel engine: A computational study *Fuel* **335** (127042)

[20] Duarte N, Arango D et.al. 2022 Experimental evaluation of low displacement compression ignited engines operating with hydroxy gas as a supplementary gaseous fuel *Heliyon* **8** (11) (e11545) pp 1-15

[21] M. Popa, N. Negurescu, C. Pana, Diesel Engines. Processes, MatrixRom, Bucharest, 2003

[22] Jung D., Assanis D.N., Modeling of Direct Injection Diesel Engine Emissions for a quasi-dimensional Multi-zone Spray Model, International Journal of Automotive Technology, volume 5, number 3, pp. 165-172, 2004

[23] Haiyan Miao, Fundamental Studies of Combustion Phenomena in Dual Fuel Engines, PhD-Thesis, School of Mechanical and Manufacturing Engineering, The University of New South Wales, 2001

Acknowledgments
The authors address special thanks to AVL GmbH Graz Austria for providing the necessary equipment's. This work was supported by a grant of the Ministry of Research, Innovation and Digitization, CCCDI - UEFISCDI, project number PN-III-P2-2.1-PED-2021-0427, within PNCDI III.

Theoretical and experimental research on the use of hydrogen in the automotive spark ignition engine

R Georgescu[1], C Pana[1], N Negurescu[1], A Cernat[1], C Nutu[2], C Sandu[1]

[1]University Politehnica of Bucharest, Department of Thermotechnics, Engines, Thermal Equipment and Refrigeration installations.
[2]University Politehnica of Bucharest, Department of Automotive Engineering.

Abstract. Reducing the level of pollutant emissions of internal combustion engines requires the adaptation of solutions that improve the combustion process in the cylinder but those solutions may increase their constructive complexity. For this reason, the use of hydrogen is an effective solution, both for replacing fossil fuels and for reducing the concentration of pollutant emissions. The use of hydrogen as an additional fuel for the automotive spark ignition engines offers the possibility of obtaining increased energy performance, decreased pollutant emissions level and decreased concentration of greenhouse gas emissions. The paper presents the results of the some theoretical and experimental investigations carried on an automotive spark ignition engine, analysing the influence of the hydrogen on combustion and implicitly on energy performance and pollution. The simulation of thermo-gas-dynamic processes in the cylinder of the engine fed successively with gasoline/gasoline and hydrogen, allowed the authors to develop a specific and efficient procedure for conducting experimental investigations. The results of the theoretical and experimental research obtained and presented in the paper highlight the advantages of using hydrogen as an additive to gasoline fuelling in the automobile spark ignition engine: the improvement of combustion (reduction of combustion duration, increase of indicated thermal efficiency, reduction of specific energy consumption) and reduction of the level of polluting emissions and of the emission of greenhouse gases. An engine adjustment strategy for the sharp reduction of the concentration of nitrogen oxides is also presented.
Keywords: pollutant emissions, efficiency, fuel consumption, simulation.

1. Introduction

In the current context of the increase in the need for energy, the price of its production and the tightening of pollution norms, discovering of new efficient solutions for fuelling the spark ignition engine is an important goal. Considering the consumption of fossil fuels and the possible exhaustion of their resources, a very good solution for fuelling the spark ignition engine is represented by non-conventional fuels. Among them, fuelling the spark ignition engine with hydrogen is very promising for increasing the effective break efficiency, reducing fuel consumption and reducing the level of pollutant emissions. Hydrogen is the most widespread element in the Universe, representing over 70% of the matter of the Universe. On Earth, it can be found in the air (in small quantities), combined with oxygen as a constituent part of water and combined with carbon as a constituent part of hydrocarbons.

One of the main advantages of using hydrogen as a fuel for spark ignition engines, besides the important advantages regarding the improvement of combustion, is represented by the various production methods, which can be both carbon free and from fossil fuels [2, 3, 4, 5]. The main methods of hydrogen production are: steam reforming of methane, partial oxidation of hydrocarbons, gasification of biomass, coal and waste, thermal cracking and, last but not least, electrolysis of water [2, 3, 4, 5].

Content from this work may be used under the terms of the Creative Commons Attribution 3.0 licence. Any further distribution of this work must maintain attribution to the author(s) and the title of the work, journal citation and DOI.
Published under licence by IOP Publishing Ltd

Most of these methods, with the exception of water hydrolysis and thermal cracking, may produce carbon dioxide.

1.1. Hydrogen properties.

The main properties of hydrogen can be found in the table 1, compared with the characteristics of gasoline and diesel fuel.

Table 1. Properties of different fuels used at internal combustion engines [1].

Property	Hydrogen	Gasoline	Diesel fuel
Chemical formula	H_2	C_8H_{18}	$C_{14,5}H_{30}$
Molar weight [kg·kmol^{-1}]	2,016	114	204
Density [kg/m^3] at 0°C, 760 mmHg	0,0899	735-760	825-870
Lower heating value [kJ/kg]	119617	42690	42600
Autoignition temperature [K]	858	753-823	473-493
Research Octane Number RON [-]	130	90-98	30
Cetane Number CN [-]	-	-	45-55
Flame speed [m/s] λ=1,20°C, 760 mmHg	2.37	0.25-0.35	-
Air-Fuel Ratio as kg air/kg fuel [-]	34.32	14.5	14.76
Minimum ignition energy [mJ]	0.018	0.2-0.3	-
Adiabatic flame temperature [K]	2300	2411	2327
Boiling temperature [K]	20.27	-	436-672
Diffusivity [cm^2/s]	0.63	-	-
Flammability limits λs.....λi	0.136-10.12	0.709-1.1	0.34-1.68
(%) $_{vol}$	4.1-75.6	1.48-2.3	0.7-5

Since the density of hydrogen is very low, hydrogen develops a tendency to quickly occupy all the space it has available, therefore it is possible that the air admitted to the engine will be quantitatively reduced, therefore the power of the engine could suffer a reduction, thus specific methods of fuelling the engine must be used, or hydrogen could be used as an additional fuel. The diffusion speed of hydrogen is approximately 7...8 times higher in gasoline vapours, which results in facilitating rapid mixing with air, even at low dosages and high engine speeds, which is an advantage for reducing the cyclic variability of the engine and increasing the stability of its operation. The boiling and critical points having very low temperatures translate into difficulties in storing hydrogen on board of the vehicle, among all the methods storing in metal hydrides being a very good choice. The combustion speed of hydrogen is higher than that of conventional hydrocarbons, therefore, when used in a spark ignition engine, the rapid combustion phase is shortened, which can have the effect of increasing the maximum pressure inside the cylinder and its rate of pressure rise. Due to the very low minimum ignition energy, there may be a very high tendency to self-ignition from hot surfaces such as the exhaust valve, spark plug, areas on the piston head, scale deposits or hot gases. The laminar burning speed, which is 7...10 times higher than that of hydrocarbons, reduces the initial phase of combustion, and the turbulent burning speed, which is approximately 2 times higher than that of hydrocarbons, can reduce the rapid burning phase. In the case of using stoichiometric mixtures, the burning speed of hydrogen may become excessive, the combustion may have knock tendencies, and a solution to avoid this phenomenon is the use of lean hydrogen-air mixtures. From the point of view of the combustion kinetics, the increase in the weight of the OH species in the combustion gases of the spark ignition engine powered by hydrogen, can be considered a precursor to the occurrence of knock [6]. Also, the intensity of the knock can be modified by other factors such as: the initial temperature of the mixture inside the cylinder [7]; coefficient of excess air lambda; replacing a quantity of air from the intake manifold by using EGR [8, 9]; compression ratio; hot spots in the combustion chamber, the degree of swirl inside the cylinder [10]. In order to reduce the heat of reaction in the case of fuelling the spark ignition engine with hydrogen, lean mixtures can be used or the ignition timing can be changed in the direction of its decrease, which moves the combustion towards expansion stroke. However, a considerable reduction of the spark advance when has negative effects on the overheating of the exhaust valves. To reduce the effect of increasing the intensity of knock due to the high temperature of the EGR, EGR cooling can be used, a phenomenon that also mitigates the effect of diluting the intake air.

Regarding the lower heating value of hydrogen, it is higher than that of conventional hydrocarbons, which is a beneficial aspect for increasing the effective efficiency of the engine and for reducing its effective specific consumption. The temperature of the adiabatic flame of hydrogen being 2318 K [11] may indicate an increase in the global temperature in the cylinder, which may have an effect on the increase in the emission of nitrogen oxides [11]. Considering the correlation between increased flame speed and the possibility of burning lean mixtures, when fuelling the spark ignition engine with the addition of hydrogen, the level of carbon monoxide emission decreases, in the cylinder the conditions for complete carbon oxidation are accomplished [11,12]. The engine's smoke emission level decreases, (especially under the conditions of using direct injection for engines that are fuelled with hydrogen in dual mode, hydrogen being injected into the intake manifold of the engine and gasoline being injected directly into the cylinder), because by increasing the speed of combustion decreases the possibility of fuel drops reaching the cold areas of the combustion chamber and thus decreases the possibility of them thermally cracking [11, 13].

1.2. Fuelling methods

The spark ignition engine can be fuelled with hydrogen by indirect injection or direct injection. The advantage of indirect injection near the valve stem is that the feeding method can be easily applied to already built engines. Also, the classic gasoline supply system can be kept and thus the engine can operate in dual mode, with gasoline through the classic system and with hydrogen through the system added in parallel. In the case of fuelling the spark ignition engine with hydrogen by the direct injection method, the injectors must be mounted in the combustion chamber, and the engine requires considerable modifications, the method lends itself especially to engines under the design phase. An important advantage of direct injection is that filling the cylinder with air is no longer affected as in the case of indirect injection. The research carried out in the works [14, 15] on a spark ignition engine equipped with an electronically controlled hydrogen indirect injection system revealed an increase in the maximum pressure in the cylinder, an increase in the speed of pressure rise and an increase in the maximum temperature in the cylinder. Due to the higher burning speed of hydrogen, a reduction in the total duration of combustion was recorded proportional to the amount of hydrogen injected, the minimum duration of combustion being recorded at the maximum amount of hydrogen used. An increase in maximum pressure was also recorded in the paper [16]. In articles [17] and [18], the authors fuelled the spark ignition engine with gasoline and hydrogen through the method of indirect injection, using dual fuelling, the gasoline being injected through the classic fuel system. According to the authors, the improvement of the combustion process was ensured by the optimal positioning of the point of maximum pressure immediately in the vicinity of the top dead centre, due to the higher burning speed of hydrogen, which increased the overall burning speed in the cylinder [17, 18].

2. In cylinder processes modelling.

In order to model the in cylinder processes, a one-dimensional model has been chosen based on Vibe formal combustion laws. The general assumption of this model is that the charge in the cylinder is considered a chemically and thermally homogeneous perfect gas mixture [1]. Combustion of fuel is assimilated with formal Vibe type laws. The calculation was carried out by the iterative method, starting from the end of the intake process, dividing the motor cycle into intervals of known size (1 degree CRA), then on each interval separately calculating the state variables. The calculation was completed when the exhaust valve was opened. The balance equation for the one-dimensional model is [1]:

$$\frac{dU}{d\alpha} = \frac{dQ_a}{d\alpha} - \frac{dL}{d\alpha} - \frac{dQ_p}{d\alpha}, \tag{2.1}$$

where:

- $\frac{dU}{d\alpha}$ is the change in internal energy;
- $\frac{dQ_a}{d\alpha}$ is the variation in the amount of heat released by combustion;
- $\frac{dL}{d\alpha}$ is the mechanical work variation;
- $\frac{dQ_p}{d\alpha}$ is the variation in the amount of heat transferred to the walls, [1].

The definition of the heat release rate characteristic was done by means of a Vibe function, [1]:

$$\frac{d\xi}{d\alpha} = 6.9\frac{(m+1)}{\alpha_s}\left(\frac{\alpha-\alpha_s}{\alpha_a}\right)^m e^{\left[-6.9\left(\frac{\alpha-a_s}{\alpha_a}\right)^{m+1}\right]} \tag{2.2}$$

where:

- m – Vibe law shape coefficient;
- α_a – combustion duration;
- α_s – start of combustion angle;
- α – current angle, [1].

To evaluate the heat transfer to the walls the Woschni 1978 model has been chosen.

2.1. Theoretical investigations results

The above presented model has been applied to simulate the in cylinder processes for a A15MF-1.5 DOHC turbocharged spark ignition engine fuelled with gasoline and hydrogen. The specifications of the engine are presented in the table 2. The chosen working regime was 2500 rpm and 50% engine load, and one case of hydrogen fuelling has been investigated, the case where the gasoline was 5% energetically substituted with hydrogen. The substitution ratio of gasoline with hydrogen is given by the following equation, [1].

$$x_c = \frac{m_{H_2}H_{i_{H_2}}}{m_{H_2}H_{i_{H_2}}+m_{gasoline}H_{i_{gasoline}}}\cdot 100 \tag{2.3}$$

Table 2. The specifications of the A15MF engine.

Number of cylinders	4
Displacement [cm³]	1498
Compression ratio [-]	9.5
Bore [mm]	76.5
Stroke [mm]	81.5
Connecting rod length [mm]	123.5
Maximum power [kW] / speed [rpm]	63/4800
Maximum torque [Nm] / speed [rpm]	136/3400

The calculated in cylinder pressure evolution is presented in figure 1. It can be observed that due to the presence of hydrogen in the combustion chamber of the engine, the maximum pressure is higher than in the standard case of fuelling only with gasoline. This can be explained by the higher flame speed of hydrogen and by its higher caloric heating value, which enhances the combustion process. The maximum rate of pressure rise also increases in the case of hydrogen fuelling, figure 3, which indicates that in the presence of hydrogen the combustion became faster. The in cylinder temperature also increases in the case of hydrogen fuelling, figure 2, because of a higher flame temperature.

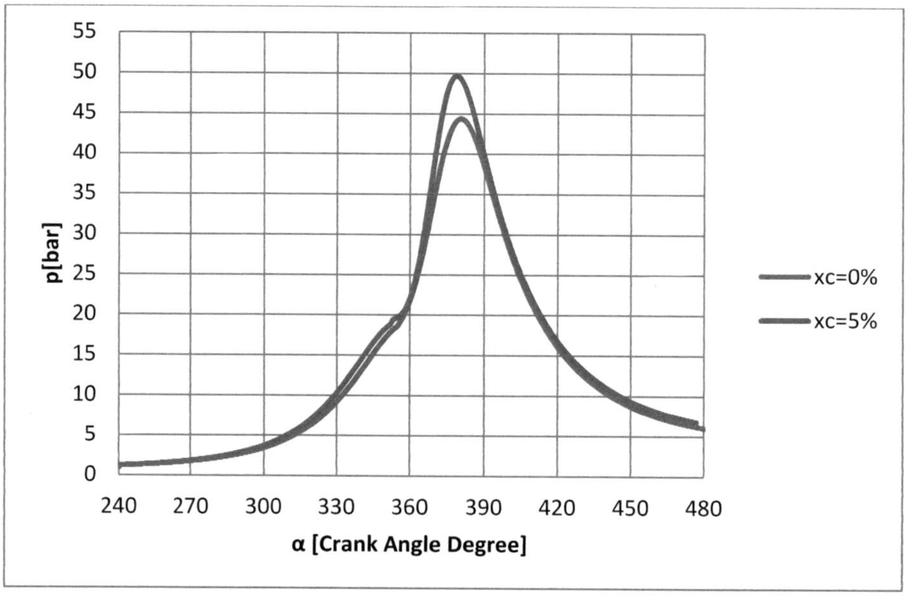

Figure 1. The calculated in cylinder pressure for the investigated cases.

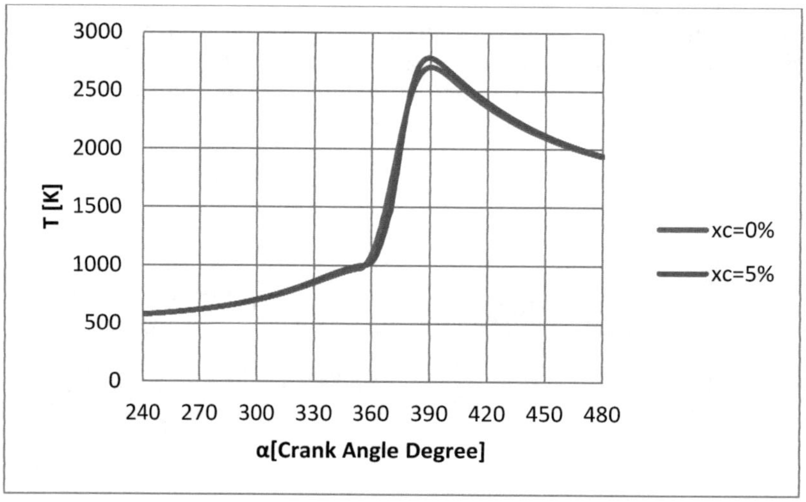

Figure 2. The calculated in cylinder temperature for the investigated cases.

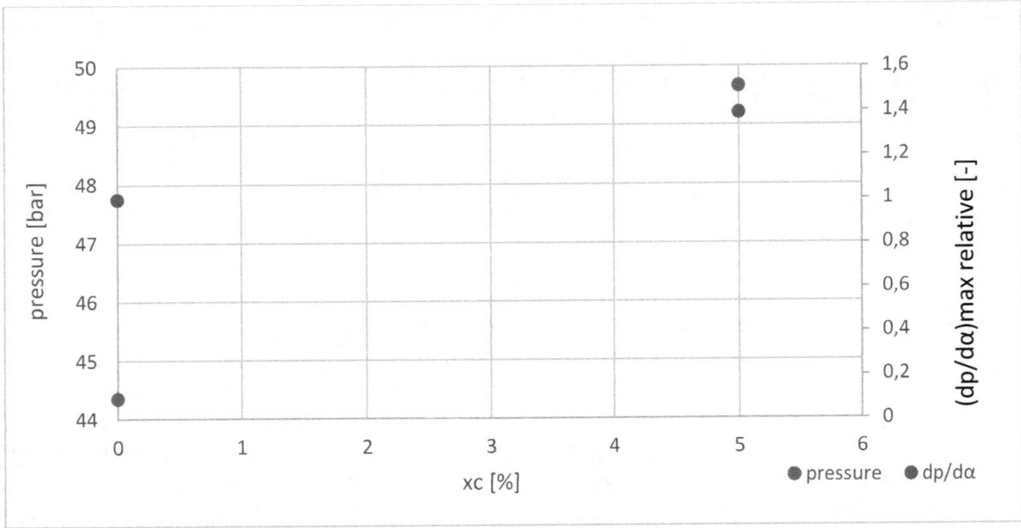

Figure 3. The pressure and the maximum pressure rise versus the substitute ratio of gasoline with hydrogen, the maximum pressure rise is presented relative to the standard case of gasoline fuelling.

3. Experimental investigations

The experimental investigations have been carried out on the same engine presented in the second subchapter, the A15MF-1.5 DOHC, with the specifications in the table 2. The test bed diagram is presented in figure 4.

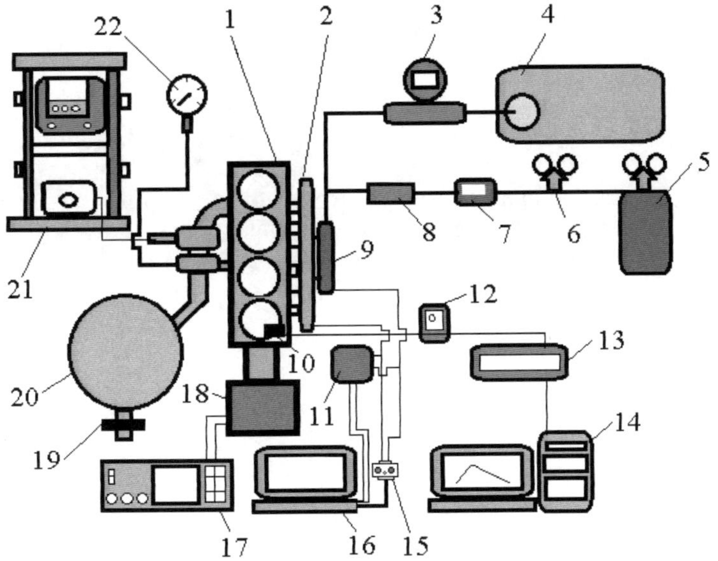

Figure 4. Schema of the experimental test bed of the SI engine fuelled with gasoline and hydrogen.

The experimental test bed, figure 4, has the following main parts:1-A15MF spark ignition engine, 2-gasoline electronic injection system,3-Krohne Optimass 3050C flowmeter for gasoline, 4-gasoline tank, 5-hydrogen tank with pressure reducer, 6-second pressure reducer, 7-Alicat Scientific MCR flowmeter for hydrogen, 8-flame arrestor, 9-hydrogen electronic injection system, 10-AVL GU12 P piezoelectric transducer for in-cylinder pressure measuring, 11-Dastek Unichip Q open electronic control unit, 12-AVL 3067 A electric charge amplifier, 13-Indimodul 621 for data acquisition, 14-PC equipped with data acquisition board Indimodul Card C1 and with AVL Indicom v.2.0 acquisition software, 15-remote for fine tune of fuels injectors control, 16-PC equipped with Dastek Unichip Program, 17-AVL dyno control cabinet, 18-AVL Dyno Perform 160 eddy current dynamometer, 19- differential pressure gauge for inlet air quantity measurement, 20-inlet air tank, 21-AVL DiCom 4000 gas analyser, 22-inlet supercharging pressure gauge.

Before starting the experimental procedure all the equipment were calibrated. The fuelling system designed for hydrogen use is the hydrogen supply system is made up of a hydrogen bottle, which has a pressure of 150 bar, this being provided with a pressure regulator that allows the adjustment of the working pressure in the range of 10-150 bar. The hydrogen bottle is connected to a mass flowmeter used to determine by direct measurement the flow of hydrogen consumed by the engine. Between the flowmeter and the hydrogen bottle, a second pressure regulator is mounted that allows the pressure in front of the flowmeter to be adjusted in the range of 10... .0.5 bar. A flame arrestor is mounted between the hydrogen flowmeter and the engine's fuelling system. The fuelling system is provided with a system of electronically controlled gas injectors, which injects gas into the intake valves port, being used for hydrogen supply. With the help of the open-type electronic control unit, Dastek unichip Q, and through the Dastek program, when opening the hydrogen injectors, the opening time of the gasoline injectors is reduced so as to maintain the same value of the engine's effective power. Through the same procedure, by reducing the opening duration of the gasoline injectors, the air-fuel mixture is leaned.

3.1. The working procedure.

The fuelling method used to fuel the spark ignition engine with hydrogen was the indirect injection, in cylinder being admitted a homogeneous mixture of hydrogen and air, and further being mixed with the gasoline-air mixture, the gasoline being injected by the standard indirect gasoline injection system. First was determined the reference, fuelling the engine only with gasoline, then the gasoline was 5% partially substituted with hydrogen, aiming to maintain the same working regime and the same power and torque as in the standard case. The engine working regime was 2500 rpm and 50% engine load. The energetic substitution ratio of gasoline with hydrogen is given by the equation (2.3).

3.2. Experimental results.

In figure 5, the maximum in cylinder pressure is presented versus the substitution ratio of gasoline with hydrogen. The maximum pressure inside the cylinder increased for the case of fuelling with hydrogen because of a higher flame speed due to the enhancement of the combustion process. The maximum pressure may increase due to the reduction of the combustion duration at hydrogen use comparative to the classic fuel use case. The maximum rate of pressure rise also increased, in figure 6 being presented relative to the standard case of fuelling only with gasoline.

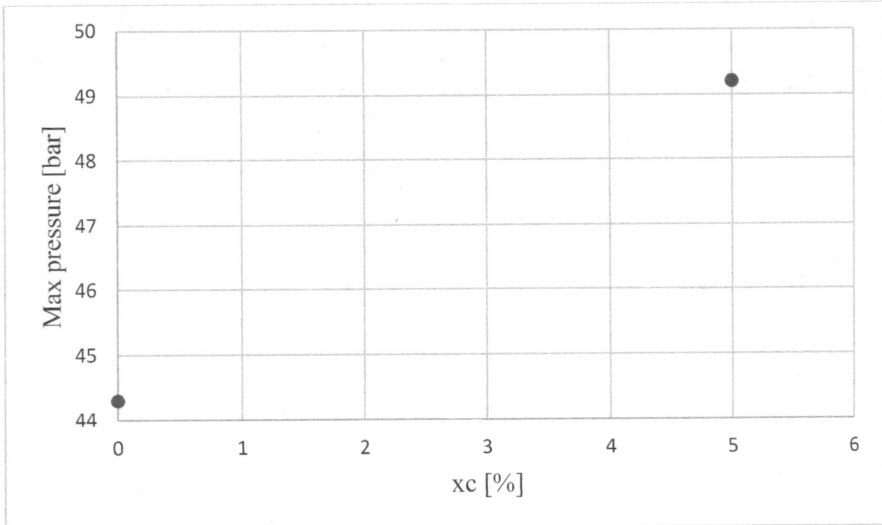

Figure 5. Maximum in cylinder pressure for the investigated cases.

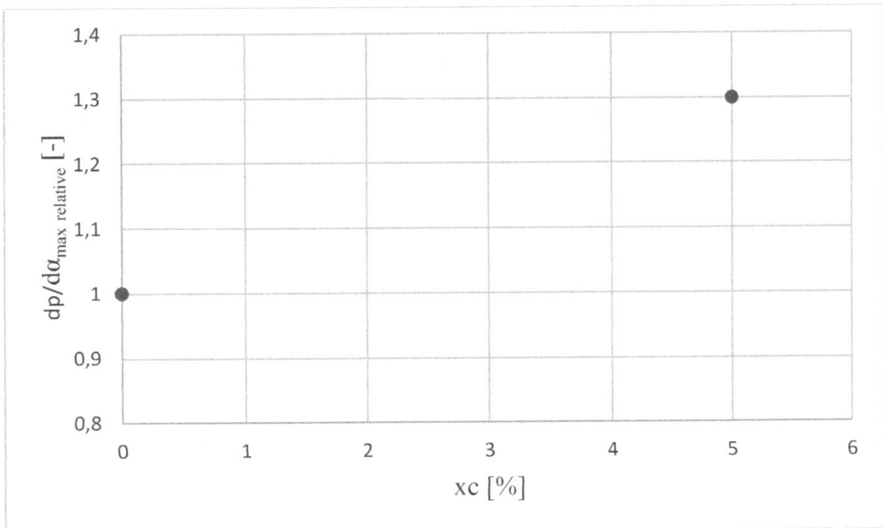

Figure 6. Maximum rate of pressure rise, relative to the standard case of fuelling only with gasoline.

The brake specific energetic fuel consumption decreased for the case of fuelling with hydrogen, due to higher value of the lower heating value of hydrogen comparative to gasoline. In figure 7 is presented the BSEC for the investigated cases.

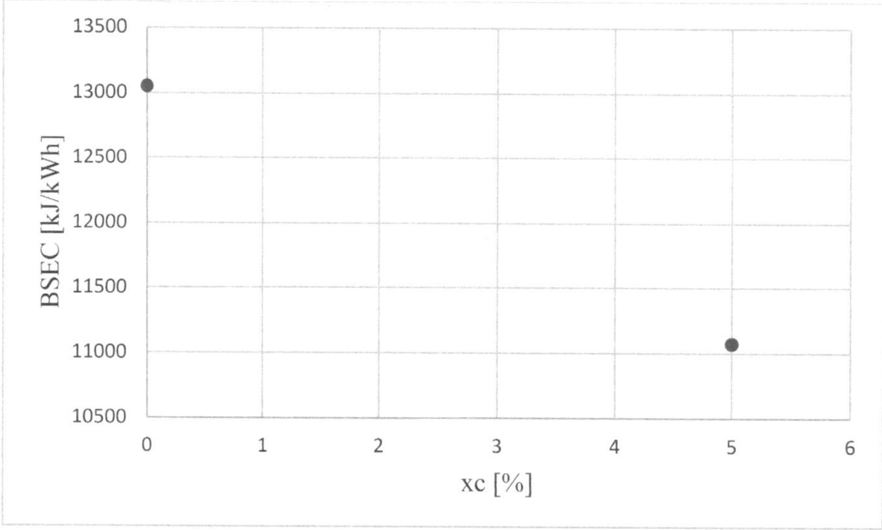

Figure 7. The brake specific energetic consumption.

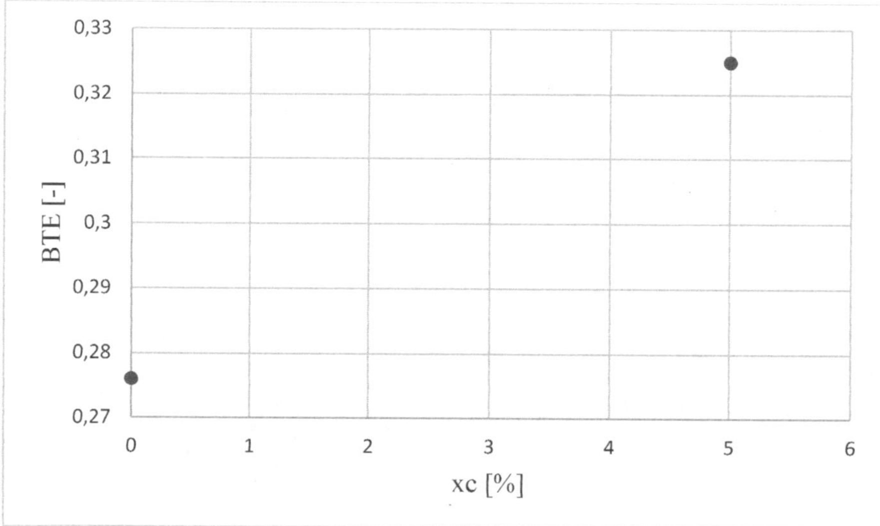

Figure. 8. The brake thermal efficiency.

The brake thermal efficiency also increased in the case of fuelling with hydrogen. In figure 8 is presented the BTE evolution for the investigated cases, relative to the case of fuelling only with gasoline.
The nitrogen oxides emission level decreases when the engine was fuelled with hydrogen, figure 9, the same trend has been obtained for the CO_2 and unburnt hydrocarbons emissions level, figures 10 and 11.

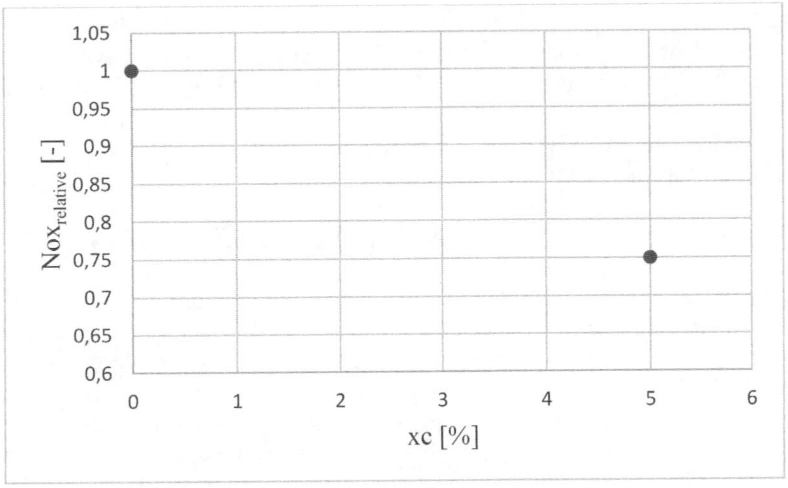

Figure 9. The relative nitrogen oxides emission level.

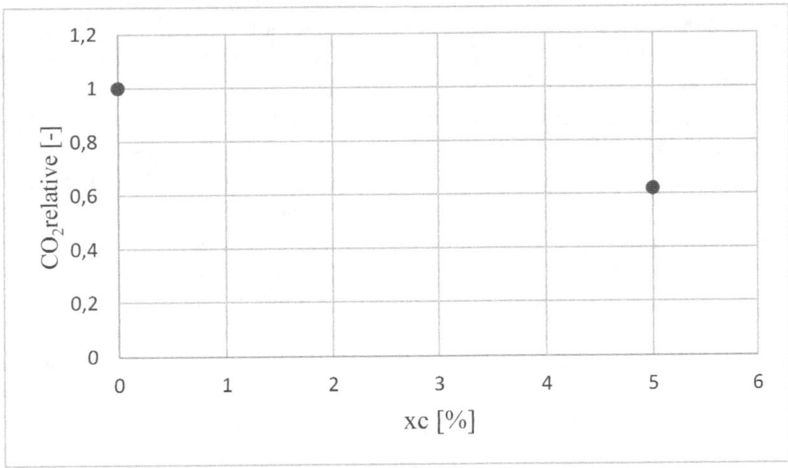

Figure 10. The relative carbon dioxide emission level.

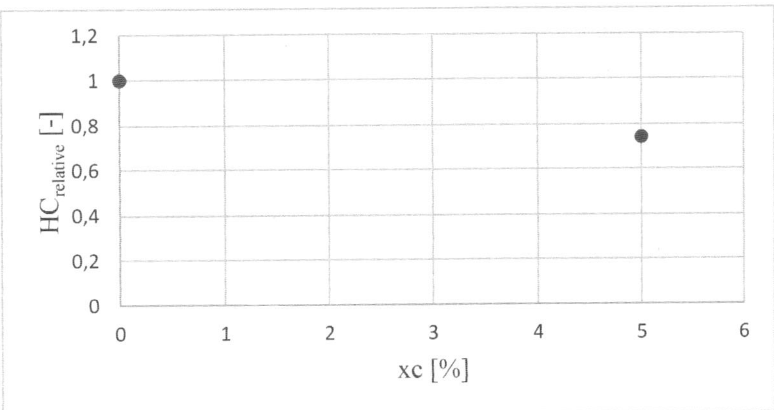

Figure 11. The relative unburnt hydrocarbons.

4. Conclusions

1. The in cylinder maximum pressure and the maximum rate of pressure rise increased when the engine was fuelled with hydrogen with 10% and 23% respectively versus classic fuelling.
2. The specific energetic fuel consumption decreased with 15%, till 11075 kJ/kWh, in the case of hydrogen fuelling comparative to gasoline fuelling.
3. The brake thermal efficiency increased with 14.4% in the case of hydrogen fuelling versus gasoline use.
4. The nitrogen oxides emission level decreased with 25% when the engine was fuelled with hydrogen comparative to classic fuelling.
5. The carbon dioxide emission level and the unburnt hydrocarbons emission level decreased with 39% and 22% respectively for the case of fuelling with hydrogen versus gasoline fuelling.
6. The operating regime investigated and analyzed in the present research, 50% engine load and speed of 2500 rpm, is a regime frequently used in the exploitation of the automotive engine in urban regime and therefore presents a high interest. The reduction of pollutant emissions, HC decreases with 22%, NO_x decreases with 25% and the GHG gases CO_2 decrease with 39% and the reduction with 15% of the specific consumption and increase of thermal efficiency with 14.4% at this regime being positive and interesting aspects, in the context of reducing the level of emissions in the urban environment. The study of the operation of the hydrogen fueled engine is extensive and covers a wide range of speeds and engine loads, including the full engine load regime and the nominal speed regime, the results being presented in future research.
7. Using substitution degrees as low as 5% presents some advantages such as: maintaining the amount of air per cycle without affecting the volumetric efficiency, limiting the increase of maximum pressure and maximum temperature during combustion to values close to the operation when fueled with gasoline, a small increase in the pressure rise rate (up till 50 MPa and 1.4 bar/crank angle degree), engine reliability is not being affected. The use of small cyclic doses of hydrogen, such as 5%, also has the advantage of maintaining the constructive simplicity of the fuelling system, whilst the use of higher degrees of substitution or only hydrogen use may imply more important constructive changes. A hydrogen fuelling system for the use of small percentages of 5% can be easily adapted to engines already in operation, improving their energetic and pollution performance, offering the advantage of keeping them in operation in the near future. A possible disadvantage of the use of small amounts of hydrogen in the intake air can be constituted by the limitation of the reduction of the level of polluting emissions (NO_x, HC) and greenhouse gases (CO_2), of the specific energetic fuel consumption, in the idea that more large cyclical amounts of hydrogen would emphasize the reduction of the level of polluting emissions and the increase of thermal efficiency.

References

[1] Popa MG, Negurescu N and Pana C *Spark ignition engines. Processes,* MatrixRom, Bucharest 2013.

[2] Wróbel K, Wróbel J, Tokarz W, Lach, J, Podsadni K and Czerwinski A 2022 Hydrogen Internal Combustion Engine Vehicles: A Review *Energies* **15** 8937.

[3] Ogden JM 1999 Prospects for Building A Hydrogen Energy Infrastructure. *Annu. Rev. Energy Environ* **24** pp 227–279.

[4] Surygała J 2008 *Wodór Jako Paliwo Wydawnictwa Naukowo-Techniczne* PWN: Warszawa, Poland.

[5] Barreto L, Makihira A and Riahi K 2003 The hydrogen economy in the 21st century: A sustainable development scenario. *Int. J. Hydrogen Energy* **28** pp 267–284.

[6] Szwaja S and Naber JD 2008 Impact of Leaning Hydrogen-Air Mixtures on Engine Combustion Knock *Journal of KONES Powertrain and Transport.* **15** No. 2.

[7] Das L M 1996 Hydrogen-Oxygen Reaction Mechanizm Hydrogen Engine Combustion, *International Journal of Hydrogen Energy* **21**, No. 8, pp. 703-715.

[8] Karim G A 2003 Hydrogen as a spark ignition engine fuel *International Journal of Hydrogen Energy* **28** pp569-577

[9] Nande A M, Szwaja S and Naber J D 2008 *Impact of EGR on Combustion Processes in a Hydrogen Fuelled SI Engine*, SAE Paper No. 2008-01-1039.
[10] Szwaja S, Bhandary K R and Naber J D 2007 Comparison of Hydrogen and Gasoline Combustion Knock in a Spark Ignition Engine *Int. J. of Hydrogen Energy* **32**/18, pp 5076-5087
[11]Aghahasani M, Gharehghani A, Mahmoudzadeh Andwari A, Mikulski M, Pesyridis A Megaritis T and Könnö J 2022 Numerical Study on Hydrogen–Gasoline Dual-Fuel Spark Ignition Engine *Processes* **10** 2249.
[12] Kamil M and Rahman MM 2015 Performance prediction of spark-ignition engine running on gasoline-hydrogen and methane-hydrogen blends. *Appl. Energy* **158** pp 556–567.
[13] Niu R, Yu X, Du Y, Xie H, Wu H and Sun Y 2016 Effect of hydrogen proportion on lean burn performance of a dual fuel SI engine using hydrogen direct-injection *Fuel* **186** pp 792–799.
[14] Ji C, Wang S and Zhangn B 2010 Combustion and emissions characteristics of a hybrid Hydrogen gasoline engine under various loads and lean conditions *Inter. J. Hydr. Ener.* **35** 5714-22.
[15] Ji C and Wang S 2009 Effect of hydrogen addition on combustion and emissions performance of a spark ignition gasoline engine at lean conditions *Inter. J. Hydr. Ener.* **34** 7823-34.
[16] Subramanian V, Mallikarjuna J and Mand Ramesh A 2007 Effect of water injection and spark timing on the nitric oxide emission and combustion parameters of a hydrogen fuelled spark ignition engine *Inter. J. Hydr. Ener.* **32** 1159-73.
[17] Wang S and Ji C 2012 Cyclic variation in a hydrogen-enriched spark-ignition gasoline engine under various operating conditions *Inter. J. Hydr. Ener.* **37** 1112-19.
[18] Kiyoshi I, Takashi S, Yasuhiro U, Kazuo Y and Toshihisa O 1997 *Investigation of cyclic variation of IMEP under lean burn operation in spark-ignition engine* SAE Paper 972-830.
[19] Neguрescu N, Pană C and Popa M G 1996 *Internal combustion engines. Processes,* Volume I, II, Matrixrom, Bucharest.

Acknowledgments
The authors address special thanks to AVL GmbH Graz Austria for providing the necessary equipment's. This work was supported by a grant of the Ministry of Research, Innovation and Digitization, CCCDI - UEFISCDI, project number PN-III-P2-2.1-PED-2021-0427, within PNCDI III. Partially, this work was supported by a grant of the Ministry of Research, Innovation and Digitization, CNCS - UEFISCDI, project number PCE 5/2022, PN-III-P4-PCE-2021-0777, within PNCDI III

IOP Conf. Series: Materials Science and Engineering 1290 (2023) 012011 doi:10.1088/1757-899X/1290/1/012011

Smoke emission analysis under normal operating conditions of two cars equipped with similar diesel engines but having different pollution standards

S M Burciu

Faculty of Engineering, Thermal Systems and Automotives Department, Dunarea de Jos University of Galati, Romania

Abstract: The present work show you a comparative analysis of smoke emissions in the case of two cars equipped with similar TDI diesel engines, but different generation in terms of the pollution standard in which they can be included, namely Euro 3 and Euro 5. The determinations were carried out at stabilized driving speeds (in the 3rd, 4th, 5th gears of the gearbox) on the chassis dynamometer VT-2M/B1- Dyno type and these experimental determinations showed the important role of the DPF in the case of car with Euro 5 pollution norme. Based on the determinations it was established a 43-75% reduction in smoke for speeds of 40-140 km/h, respectively a 50-80% reduction in smoke for speeds of 100-200 km/h (depending on the gear ratio used) in case of engine with DPF, that are important reductions for smoke emision in case of the speeds specific to the traffic in cities but also for the high speeds specific to the traffic outside of the cities.

1. Introduction

For the cars with diesel engines, according to the current anti-pollution legislation, it is important to measure the degree of smoke, the concentrations of nitrogen oxides, hydrocarbons and carbon dioxide emissions [3], [4], [6], [7], [17]. The composition of the exhaust gases depend on: engine speed and load, air/fuel ratio, engine wear, fuel system adjustment/ deregulation, turbocharger technical condition and of course the anti-pollution norme in which is included the car [6], [7], [8]. All new vehicles equipped with a diesel engine have been required to significantly reduce their smoke emissions with the entry into force of the Euro 6 anti-pollution standard, that meens the smoke emissions are reduced to be limited to 5 mg/km (meaning an additional 80% reduction compared to the Euro 4 standard) [3], [5], [14], [16] and this substantial reduction of smoke was made possible by the introduction of the DPF (diesel particle filter) starting with the Euro 5 and even Euro 4 standards for engines with a cylinder capacity greater than 1.7 liters.

The increasing number of passenger automobiles with diesel engines (especially second-hand cars) and the intensive traffic in urban areas have a negatively impact both the environment and human health. Smoke emissions (as indicated also in this study) depende of speed and load engine, the age and technical condition of the vehicle, speed and of course the pollution norme in which is included the car [15], [16].

In the specialized literature there are models for the formation of soot in the case of the diesel engine, among which we mention the ones presented in, [16], [17], [18].

Corresponding Email: Mugurel.Burciu@ugal.ro

Content from this work may be used under the terms of the Creative Commons Attribution 3.0 licence. Any further distribution of this work must maintain attribution to the author(s) and the title of the work, journal citation and DOI.
Published under licence by IOP Publishing Ltd

The smoke from the exhaust gases is the result of two phenomena that act differently, namely the formation in the very rich zone of the mixture at high temperatures, respectively the post-oxidation in the presence of oxygen, [16].

The soot particle comes from hydrocarbons with 10-25 carbon atoms and an approximately double number of hydrogen atoms, [16]. A complex pyrolysis process that corresponds to dehydrogenation, cracking, polymerization and condensation reactions contributes to the formation of soot.

Soot is formed following two distinct stages, namely, the formation of soot nuclei (precursor formation phase) and the actual formation of soot particles, [15], [16], [19].

The pre-injection method used in the direct injection diesel engine to reduce operating noise can also have a major disadvantage in the sense that it reduces the time available for the homogenization of the air-diesel mixture, which will lead to many areas with high enrichment in the combustion chamber, fact that will lead to an increase in smoke emission [16]. The use of post injection has the benefit of increasing the temperature at the end of combustion, which will allow the reduction of soot due to better post oxidation and the improvement of DPF efficiency [16].

For a good operation of the DPF, it is necessary to use diesel fuel with a sulfur concentration of less than 500 ppm. In these conditions, the DPF manages to retain more than 90% of the smoke particles emitted by the engine. In the first phase, the DPF retains the particles permanently, and after reaching an uploaded hight level with smoke, regeneration will be triggered, which is usually active regeneration, which involves either fuel injection upstream of the DPF or post injection into the combustion chamber commanded by the ECU, [16], [19]. According to [16], diesel additives with a cleaning effect have an important effect in reducing the emission of particles in the exhaust gases.

An important problem for the DPF are the oil particles that settle inside the DPF. Since these particles (which are non-combustible) do not burn once the soot in the regeneration process ends up clogging the DPF and removing it from use [16], [9]. For this reason, it is important that in the case of diesel engines with DPF, special oils with low sulfur content and that do not generate deposits and ash are used. Loading the DPF with oil particles will lead to more frequent regenerations and implicitly to increased diesel consumption. Of course, the technical condition of the engine (compression, distribution system and turbocharger) have a major influence on the loading of the DPF with non-combustible oil particles.

According to the experimental results of this paper, in order to reduce the smoke emission in the case of cars equipped with diesel engines, it is recommended to purchase vehicles with a minimum pollution rate of 5 or 6 and to maintain the anti-pollution system in an appropriate technical condition. Also to reduce the smoke emission in the case of cars equipped with diesel engines, it is recommended to adapt the driving speed and the transmission ratio so that, as much as possible during most of the driving time, the engine runs in the economic zone which corresponds to the range 0.55-0.65 from the engine rotation of maximum power, function area with the lowest fuel consumption [20]. If these recommendations are followed, then the smoke emission of cars powered by diesel engines would be at low, and in the case where the engine is equipped with the particle filter, the smoke emission will be substantially reduced (compared to that of the engine without DPF), according to what is presented in this work.

The results presented in the paper only refer to the cars that are running on horizontal roads and these results show a decrease of 43% to 80% of opacity index (depending on the gear ratio used and the speed) in case of engine with DPF, that are important reductions for smoke emision and this reduction certainly representing an important one in all the activity directed in the direction of the substantial reduction of the polluting emissions of thermal engines.

2. Equipment used for experimental studies

The experimental determinations were performed using the Volkswagen Passat (140HP/4200 rpm, 2011), Euro 5 pollution norme, car that is equipped with a particulate filter (DPF) and Seat Toledo II (110 HP/ 4150 rpm, 2004), Euro 3 pollution norme, car that is not equipped with a particulate filter (DPF), cars whose characteristics are indicated in the table 1. The load determinations were performed

using the chassis dynamometer VT-2M/B1- Dyno type (figure 1), [1], using the road method and for measuring absorbtion coefficient (opacity index), the Ultima 600 Johnonson Controls Automotive Electronic opacimeter was used, with the intervals and accuracy indicated in [2].
To perform the load tests, the "road test mode" was used. This mode simulates real road driving conditions for cars. Figure 2 depicts the data collecting system that comes standard with the V-TechDyno VT-2/B1 Chassis dynamometer.

Table 1. Technical specifications of cars

Technical specifications	Seat Toledo II-2004, pollution norme Euro 3, TDI with intercooler, [10], [11]	Volkswagen Passat B7 - 2011, pollution norme Euro 5, TDI with intercooler and DPF, [12], [13]
Total Max. Power (hP)/Rpm	110/ 4150	140/ 4200
Max. Torque (Nm)/Rpm	235/ 1900	320/ 1750
Engine Capacity (cm^3)	1896	1968
Power/ liter (hP/l)	58	71
Nr. cylinders	4	4
Valves Per Cylinder	2	4
Bore and Stroke (mm)	79,5 x 95,5	81 x 95,5
Compression Ratio	19,5	16,5
Max. speed (km/h)	193	213
CO_2 Emissions (g/km)	135	119
Manual transmission (1st gear to last gear)	3,78; 2,06; 1,35; 0,97; 0,77 final drive: 3,39 2600 rpm at 120 km/h	3,77; 1,96; 1,26; 0,87; 0,86; 0,72 final drive: 2,76 1950 rpm at 120 km/h
Acceleration 0-100 Km/h (s)	11,2	9,8
Combined Consumption (l/100km)	5,0	4,6
Max. Permissible Mass (kg)	1820	2100

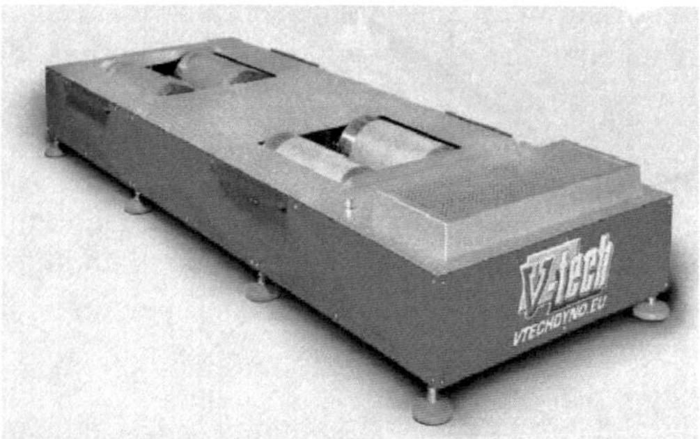

Figure 1. Chassis dynamometer VT-2M/B1-Dyno

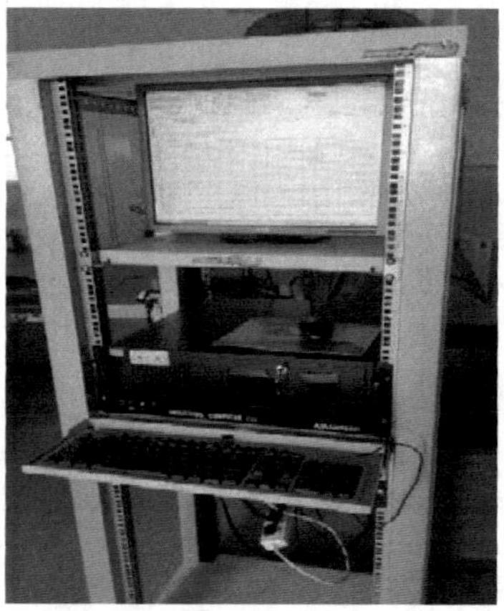

Figure 2. The V-TechDyno VT-2/B1 data acquisition

3. The results obtained from the experimental determinations

Using the indicated equipment, experimental determinations for measure the opacity index (absorbtion coefficient) at load regimes were performed, i.e. the determination of opacity index based on driving speeds. Determinations were performed at various stabilized driving speeds, specific to the movement of the car on an urban and extra-urban road, using the indicated cars (Table 1) and equipment [1], [2]. The determinations were performed considering the running of the cars on a horizontaly road.

The evolution of the opacity index, for quasi-stationary regimes (quasi-stationary regimes close to stationary regimes, which means stabilized driving speeds for cars) depending on driving speeds, and the selected gearbox ratio are shown in figures 4, 5, 6 for speeds between 40 [km/h] and 200 [km/h] driving in the 3th , 4th and 5th gear of the manual gearbox.

The variation of the opacity index depending on engine speeds is shown in figure 3.

On the evolution of the absorbtion coefficient, measured at stabilized driving speeds for various speeds and loads, according to the variations presented, the following can be indicated:

-- for engine speeds between 1500-3000 rpm (which corresponds to speeds of 40-90 km/h for the 3rd gear; speeds of 50-120 km/h for the 4th gear, respectively 65-140 km/h for the 5th gear) the reduction in opacity index is from 43% to 75% in the case of the vehicle falling within the Euro 5 pollution standard compared with the car included in Euro 3 pollution norme, with a greater reduction at a speed of around 2000 rpm (which represents the economic engine speed), that means speeds in the ranges of 50-65 km/h, 70-85 km/h, 85-95 depending on the gear box ratio;

-- for engine speeds between 3000-4350 rpm (which corresponds to speeds of 100-180 km/h for the 4th gear, respectively speeds of 130-205 km/h for the 5th gear) the reduction the opacity index is 50% to 80% in the case of the vehicle with the Euro 5 pollution standard compared with the car included in Euro 3 pollution norme, with a greater reduction at engine speeds from 3800-4000 rpm, that means speeds between 140-165 km/h, respectively 175-190 km /h, depending on the gear box ratio;

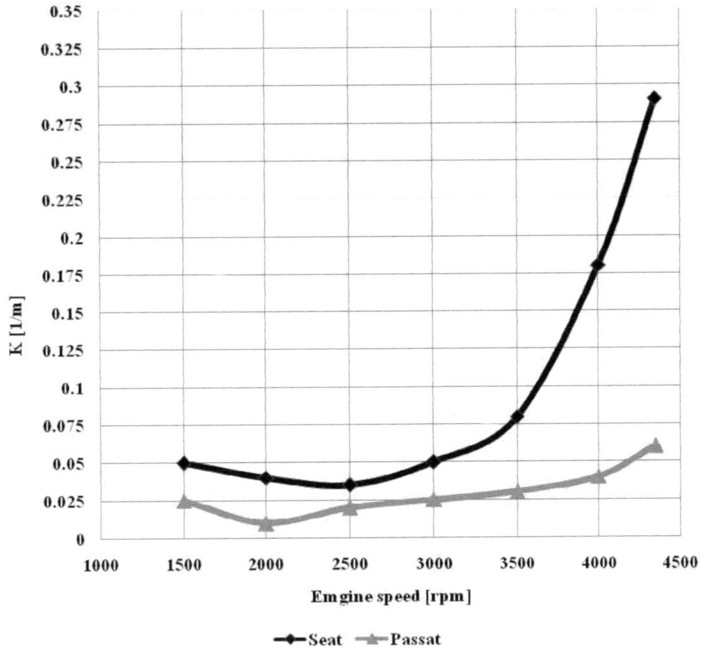

Figure 3. Absorption coefficient depends on engine speeds

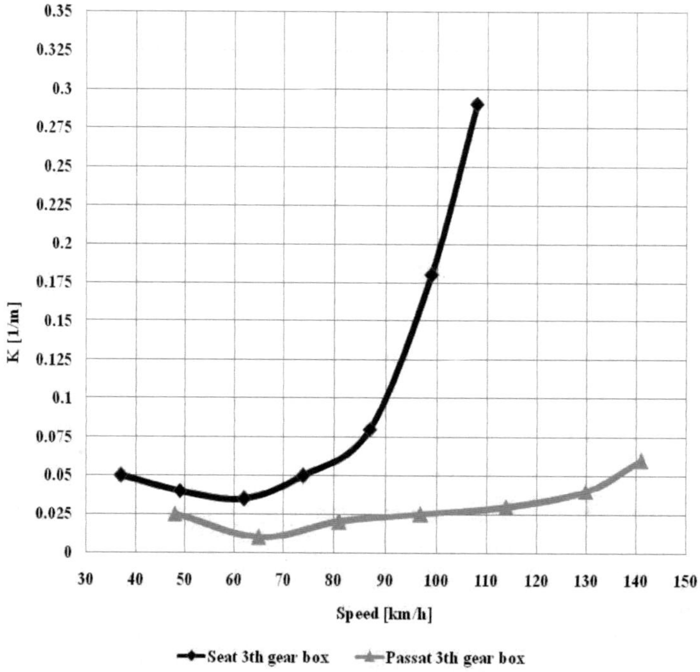

Figure 4. Opacity index depends on car speeds in 3th gear box

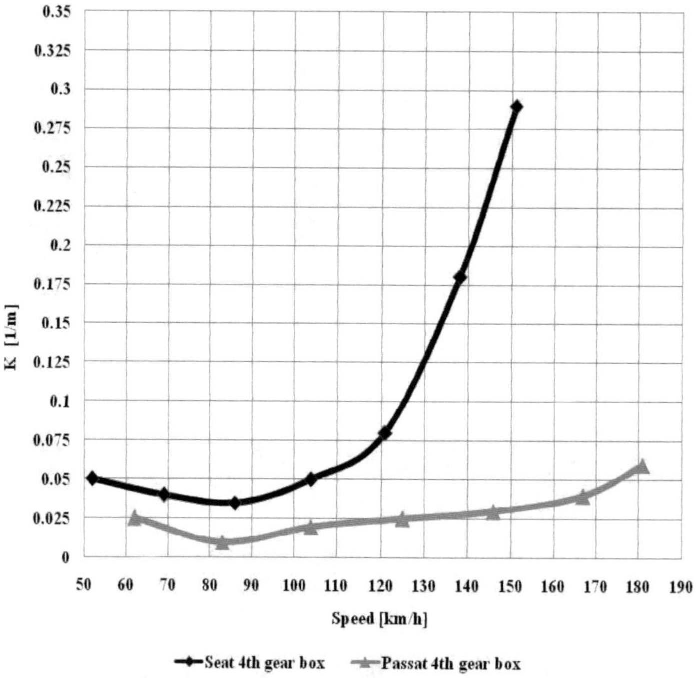

Figure 5. Opacity index depends on car speeds in 4th gear box

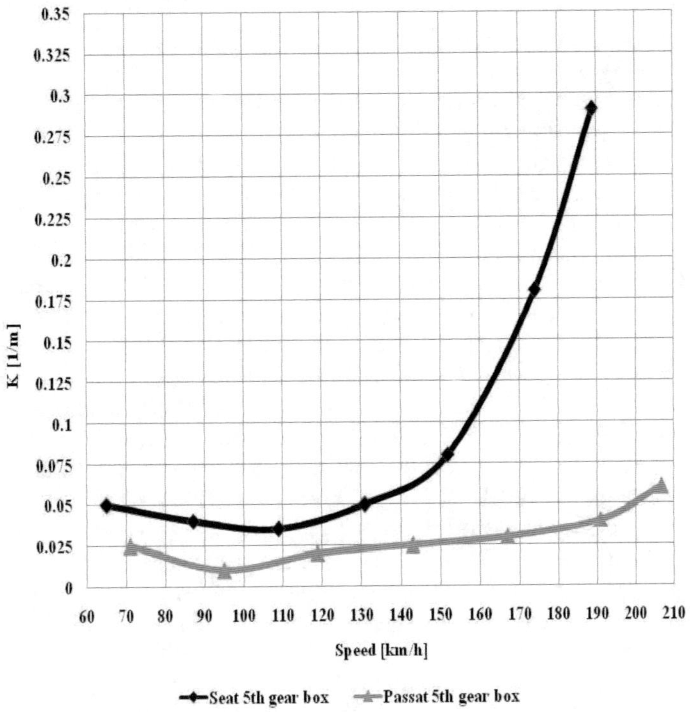

Figure 6. Opacity index depends on car speeds in 5th gear box

-- at high engine speeds (over 3000 rpm) and speeds over 100[km/h], the efficiency and role of the DPF in reducing smoke emissions in Euro 5 cars compared to Euro 3 cars can be very clearly observed, because DPF determines a reduction in smoke emission of up to 80%.

4. Conclusions

Based on the measurement of the absorption coefficient for smoke for the two cars with similar diesel engines, but with different pollution standards, it can be concluded that at low speed regimes of 40-80 km/h, specific to the urban regime, are important reductions for smoke emissions for cars with Euro 5 compared to Euro 3, this reduction being up to 75%, and at high engine speeds (over 3000 rpm) and speeds cars over 100 [km/h] we can clearly see the efficiency and the role of DPF in reducing smoke emissions in Euro 5 cars compared to Euro 3 cars, the reduction being up to 80%.

The determinations were made at different but stabilized running speeds for cars, and for this reason the smoke emission values are reduced (since there are no fast acceleration regimes). Certainly, the DPF efficiency is also high in the case of regimes with sudden accelerations (dynamic speed regimes), situations in which the smoke emission is at higher values than in the case of stabilized speeds, but such research will be carried out in the future.

In order to reduce the emission of smoke both in urban and extra-urban conditions, it is recommended to modernize the car fleet in the sense of purchasing vehicles with Euro 5 and Euro 6 diesel engines, abandoning those falling under lower pollution standards and driving in such a way that, regardless of the ratio transmission set, the engine should operate in its economic zone (in general for the diesel engine this economic zone is 1700-2500 rpm) and, of course, very dynamic regimes should be avoided (those with rapid changes in engine speed and vehicle speed). In this way, emissions will be reduced by reducing consumption, and on top of that, smoke reduction will be added using passive methods such as DPF.

IOP Conf. Series: Materials Science and Engineering 1290 (2023) 012011 doi:10.1088/1757-899X/1290/1/012011

5. References

[1] https://vtechdyno.eu/vt-2b1-chassis-dynamometer.html
[2] http://www.savbass.com/Ultima/281604780E.pdf
[3] ***, RNTR 1, https://www.rarom.ro/cs-uploads/RNTR%201.pdf
[4] ***, RAR, http://www.rarom.ro, https://www.rarom.ro/cs-uploads/listalegislatie.pdf
[5] S M Burciu – Laboratory work: Part I-Car dynamics. Part II-Car diagnostics, GUP Galati 2017
[6] S M Burciu, G Coman, Experimental study of NO_x and CO_2 emissions at load regimes for a passenger car powered by direct injection diesel engine, https://iopscience.iop.org/article/10.1088/1757-899X/595/1/012035, 2019
[7] S M Burciu, G Coman, A comparative study of smoke emissions between idling regimes and high load regimes for a standard diesel direct injection car, https://iopscience.iop.org/article/10.1088/1757-899X/444/7/072018, 2018
[8] G Coman, Mugurel Salvadore Burciu, Nicușor Baroiu, Vehicles emissions under different driving conditions in urban areas, *Revista de Chimie*, 2019, ISSN: 0034-7752
[9] Essig G, Kamp H, Wacker E. Diesel engine emissions reduction - The benefits of low oil consumption design, *International Congress & Exposition*, Detroit, Michigan, USA, February 26–March 2, 1990:1-10. SAE900591
[10] https://www.cars-data.com/en/seat-toledo-1-9-tdi-110hp-executive-specs/44539/tech
[11] https://www.auto-data.net/ro/seat-toledo-ii-1m2-1.9-tdi-110hp-13543
[12] https://www.cars-data.com/en/volkswagen-passat-2-0-tdi-140hp-bmt-highline-specs/52039
[13] https://www.auto-data.net/ro/volkswagen-passat-b7-2.0-tdi-140hp-16828
[14] ***, RNTR 11, https://www.rarom.ro/cs-uploads/RNTR11.pdf
[15] Busoi A., Ivan F., Cristea D., Experimental research concerning the validation of a DPF constructive variant for a C.I.E. meant for light drive, *Buletin Stiintific Universitatea din Pitesti, Seria Autovehicule Rutiere*, ISSN: 1453-1100, 2010.
[16] Ivan F., Lita D., Busoi A., Metode si mijloace de depoluare a motoarelor pentru automobile, *Editura Matrix ROM*, Bucuresti 2014
[17] Negrea D., Sandu V., Combaterea poluarii mediului in transporturile rutiere, *Editura Tehnica Bucuresti* 2000
[18] Sung N., Lee S., Kim H., A numerical study on soot formation and oxidation for a direct injection diesel engine, *Proceedings of the Institution of Mechanical Engineers, Part D: Journal of Automobile Engineering* 2003
[19] Lita D., Ivan F., Ways of improving the particle filters' performances used in depolluting the quick Diesel engines, *International Congress on Automotive and Transport Engineering, CONAT Brasov, Volumul 2*, ISSN 2069-0401, Brasov 2010
[20] S M Burciu and G Coman , The study of the influence of speed regimes on smoke emissions for a diesel injection car, https://iopscience.iop.org/article/10.1088/1757-899X/1262/1/012061

IOP Conf. Series: Materials Science and Engineering 1290 (2023) 012012 doi:10.1088/1757-899X/1290/1/012012

Analysis of the behaviour for a steering rack under the influence of the road elements

S Petrovan[1], A G Popa[1], B Spătaru[1], E Rakosi[1*], T Ulian[1*] and Gh Manolache[1*]

[1] Automotive and Mechanical Engineering Department, "Gheorghe Asachi" Technical University of Iași, 61-63 D. Mangeron Blvd., Iași, 700050, România

E-mail: stefan.petrovan@student.tuiasi.ro

Abstract. The automotive industry is dealing with rapid innovation and competition, which calls for either changing current products or replacing them with more modern materials. Automobile manufacturers are currently focusing on reducing the weight in order to conserve natural resources and energy. As luxury and safety equipment is added, the weight of the vehicle increases, which reduces overall performance and fuel efficiency. As a result, modern car industry leaders prioritize vehicle armoring. The use of composite materials based on plastic fibers in automobile manufacture has a number of financial and practical benefits for both the buyer and the manufacturer. Benefits are attributed to a decrease in weight, improved production standards, stylistic advancements, corrosion and abrasion resistance. Studying how various composite material types used to build directional cases behave under impact with a potentially hazardous object while moving at a speed of 100 km/h is the proposal of this article.

1. Introduction

If the term "composite" is used at its nominal value, it might be used to describe almost everything because, according to a more thorough inspection, each material is composed of various subunits. With all of them, the expression "material matrix" refers primarily to a material that has been reinforced with fibers in contemporary material engineering. For instance, the name "FRP" (Fiber Reinforced Plastic) typically refers to a matrix of polyester that contains fiberglass, and this particular composition accounts for the majority of market share in the industrial sector today [1].

The concept of "composite" wasn't invented by anyone. Wood is a type of natural material made of cellulose fibers with high stiffness and resistance incorporated into a porous, polizaharid-based matrix. Although man was able to recognize the fact, the best way to get around the two main limitations of the natural limulus-those of size (a tree has a limited transverse dimension) and anisotropy (properties are visible differently in axial and radial directions) was to create the composite material known as plywood. With all of them, nature does a much better job of designing and creating than we do [2].

Although the extent to which these materials are composed may be unnoticed, they broaden the horizons of designers across all fields of engineering.

Laminate composites have unique resistance and increased rigidity, making them a viable option for reducing the difficulty of subpar constructions. Although they have superior planned performance, these materials have the characteristics of large slabs, and the results of impact loads may include matrix fracture, fiber defect, and/or delamination [3].

Advanced materials, developed in the 1960s and first commercially released by DuPont in the 1970s under the brand name Kevlar, greatly benefit from aramid fiber armor. The higher degree of durability, high

Content from this work may be used under the terms of the Creative Commons Attribution 3.0 licence. Any further distribution of this work must maintain attribution to the author(s) and the title of the work, journal citation and DOI.
Published under licence by IOP Publishing Ltd

IOP Conf. Series: Materials Science and Engineering 1290 (2023) 012012 doi:10.1088/1757-899X/1290/1/012012

tolerance for deterioration and defective aramids contribute to good impact and ballistic. As opposed to carbon or glass fibers, aramid fibers do not exhibit brittleness when being sparged. In contrast, a small number of minor defects in the fibrils, which are molecular fires that align each aramid fiber in the same direction as its sister fiber, cause aramid fiber decomposition [3].

The company DuPont also introduced polymer polyacrylonitrile (PAN) in 1950, and it is still the only precursor that can be used to produce carbon fibers with additional resistance today. These PAN-based carbon fibers are mostly used in automobiles but as well have substantial structural uses in the aerospace industry (such as the Boeing 787). They are also used to create energy (oil lamp filaments) [4].

The carbon fiber IM7, which is used for long scarves in the aerospace industry, has a tensile strength of 5.6 GPa and a Young's modulus of 276 GPa. Recently, commercially available carbon fiber based on PAN (T1100G) with 6,6 GPa and 324 GPa became available. In a city that has been developing for more than 50 years, resistance to bending and the carbon fiber modulus based on PAN are still only about 10% and 30% of the theoretical resistance of carbon-carbon ties, respectively. On the other hand, continuous carbon fiber on a pass-based manufacturing process is possible with a modulus of up to 965 GPa (more than 90% of the theoretical modulus) [4].

The article studies what would happen if we replace steering rack housing aluminum alloy with three composite materials. The material used for reference for the simulation is LM 24 aluminum alloy (Al-Si8Cu3.5) which is currently used for construction of the case for the steering racks of the cars.

We propose the change of this aluminum with three composite materials:

- Standard carbon fiber
- High modulus carbon fiber
- Aramid fiber (Kevlar)

Multiple studies [5], [6], [7], [8] studied impact behavior and dampening capabilities of composite materials. In this article, we set up a random vibration simulation and an impact simulation on a housing for a steering rack, comparing the results with the base material.

2. CAD Data and Simulation Setup

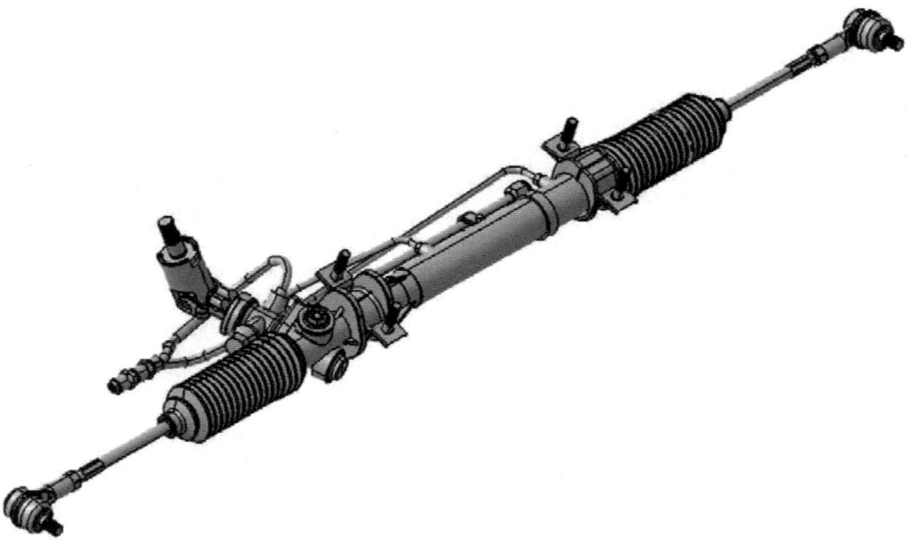

Figure 1. Standard Hydraulic steering rack used in the study.

IOP Conf. Series: Materials Science and Engineering 1290 (2023) 012012 doi:10.1088/1757-899X/1290/1/012012

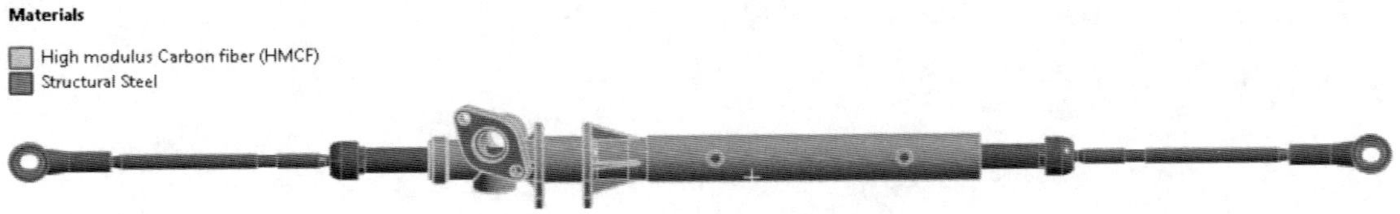

Figure 2. Material setup for the steering rack.

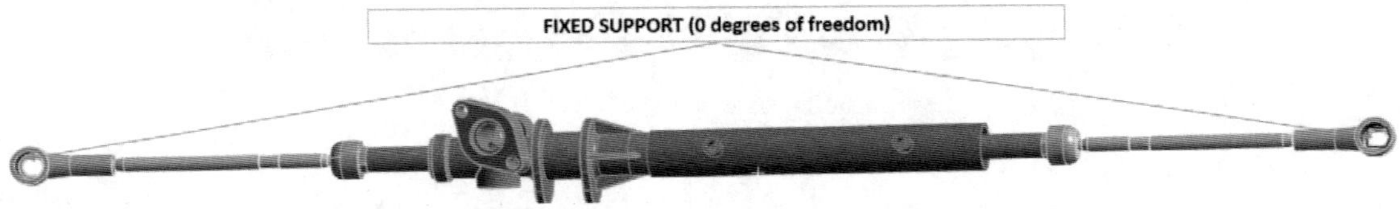

Figure 3. Boundary conditions for the simulation.

For a dynamic analysis in Ansys all the contacts are linear. In our case, all contacts between parts are bonded (fixed). This will create a worst-case scenario.

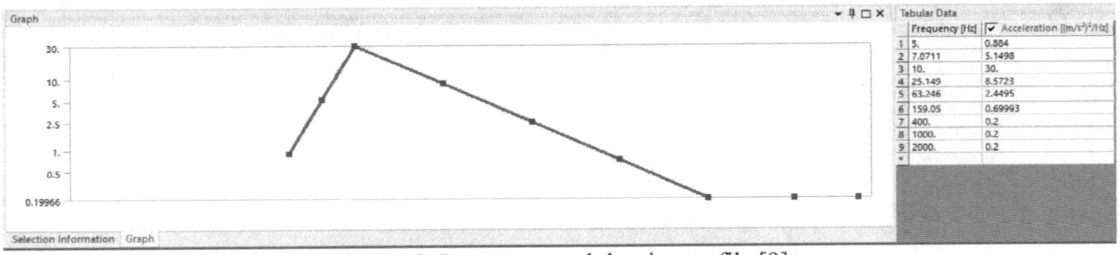

Figure 4. Power spectral density profile [9].

The measurement of a signal's power content vs frequency is called a power spectral density (PSD). Typically, a PSD is used to describe broadband random signals. The spectral resolution used to digitize the signal normalizes the PSD's amplitude. A PSD has amplitude units of Hz for vibration data. The profile is only applied in the Z direction. Excitations on other axis does not make sense due to the model's symmetry.

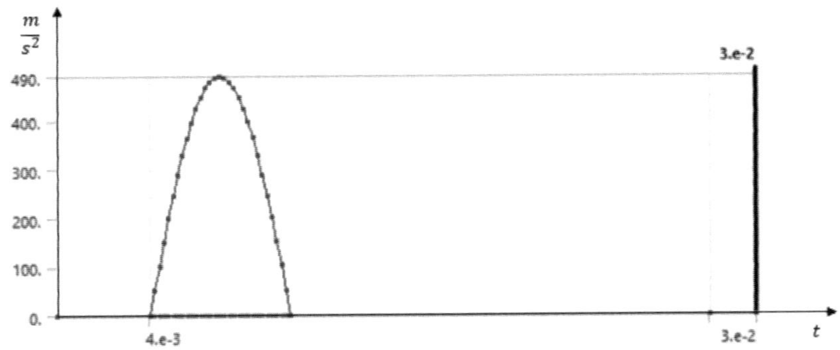

Figure 5. Shock definition.

For the shock simulation, according to IEC 60068 [10], we applied an acceleration of 50G for a period of time equal to 6 milliseconds.

IOP Conf. Series: Materials Science and Engineering 1290 (2023) 012012 doi:10.1088/1757-899X/1290/1/012012

3. Simulation of the reference material (Al-Si8Cu3.5)

3.1 Random vibration simulation

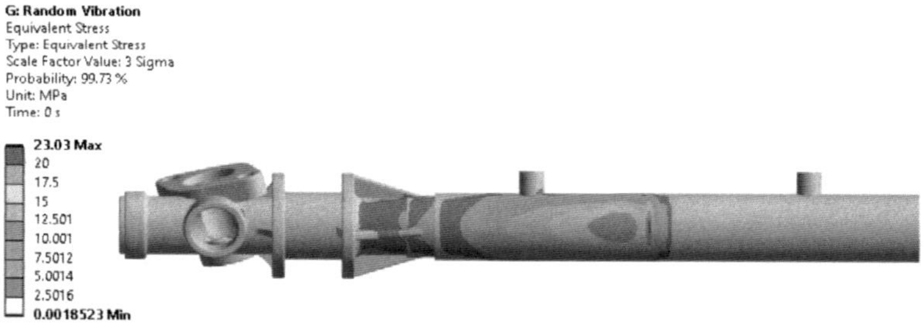

(a)

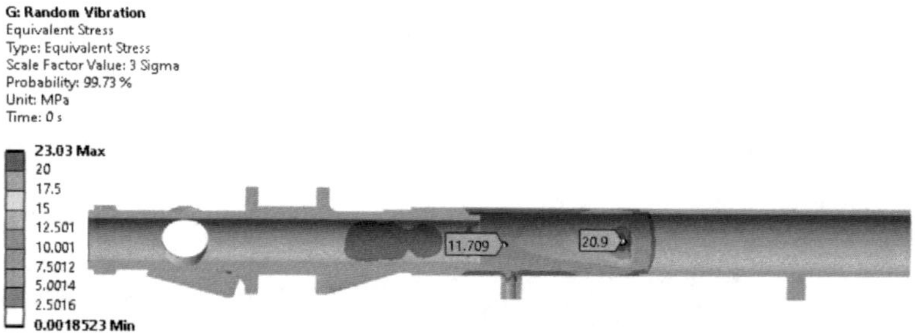

(b)

Figure 6. Equivalent stress for random vibration simulation: a) Full view; b) Section view.

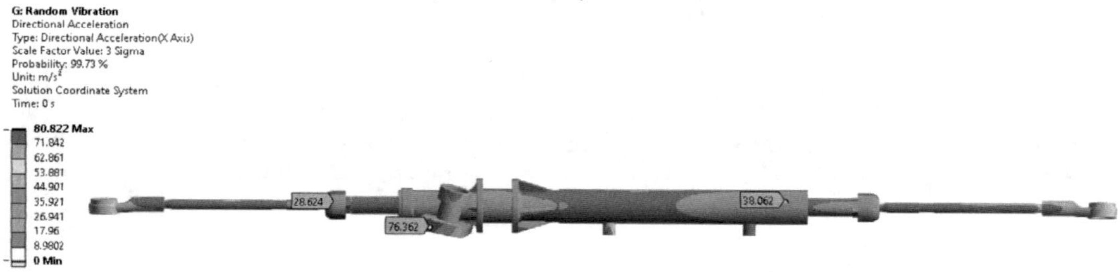

Figure 7. Acceleration response for random vibrations along the X axis.

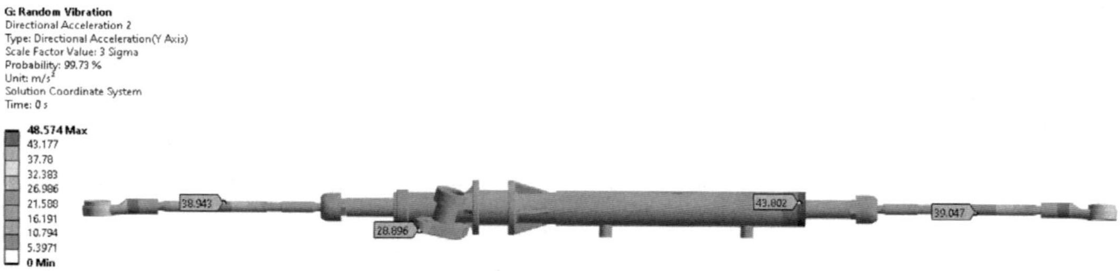

Figure 8. Acceleration response for random vibrations along the Y axis.

IOP Conf. Series: Materials Science and Engineering 1290 (2023) 012012 doi:10.1088/1757-899X/1290/1/012012

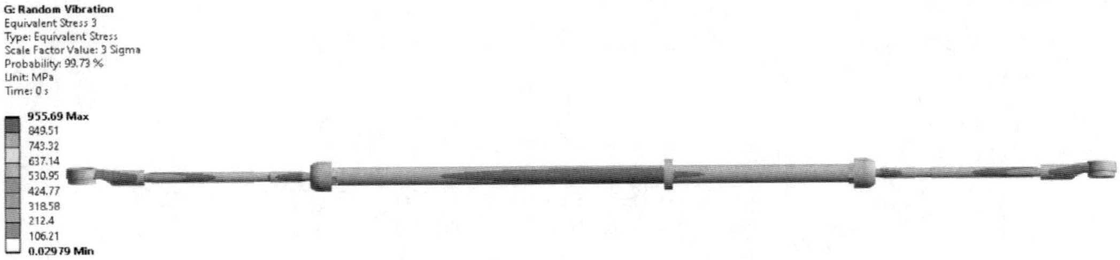

(a)

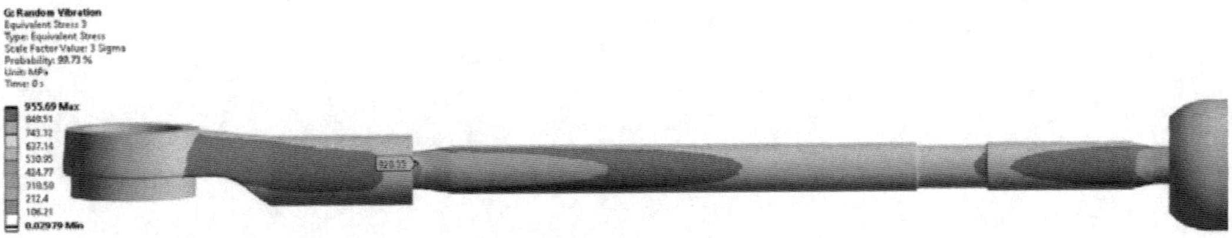

(b)

Figure 9. Random vibration simulation equivalent stress on the steering rack: a) Propagation on the whole steering rack; b) Most solicitated area.

3.2 Impact simulation

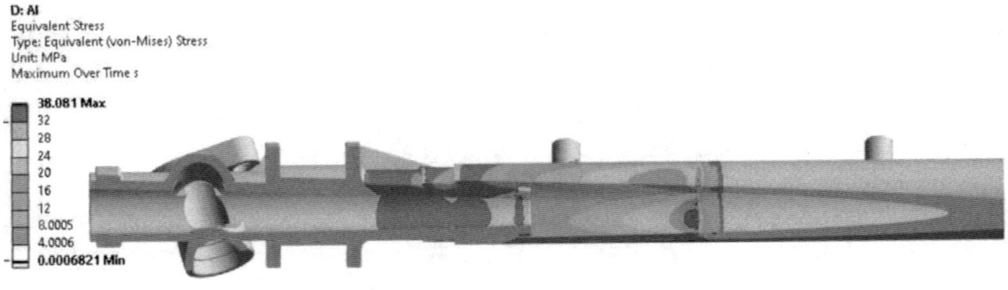

Figure 10. Equivalent stress of the steering housing.

4. Aramid fiber (Kevlar)

4.1 Random vibration simulation

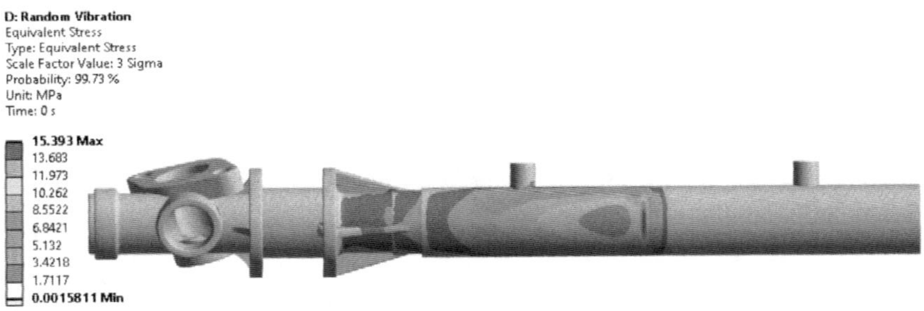

(a)

IOP Conf. Series: Materials Science and Engineering 1290 (2023) 012012 doi:10.1088/1757-899X/1290/1/012012

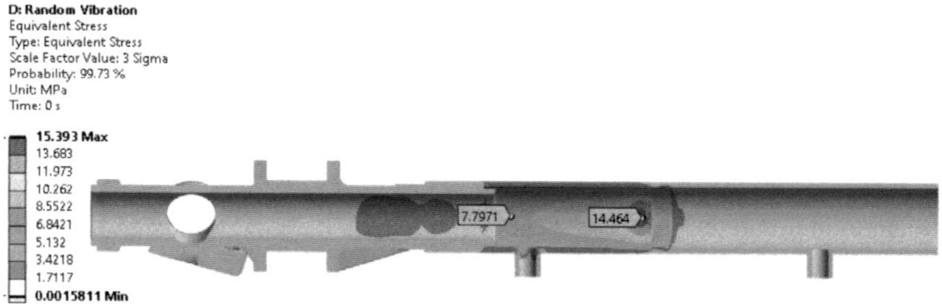

(b)

Figure 11. Random vibration simulation equivalent stress:
a) Full view; b) Section view.

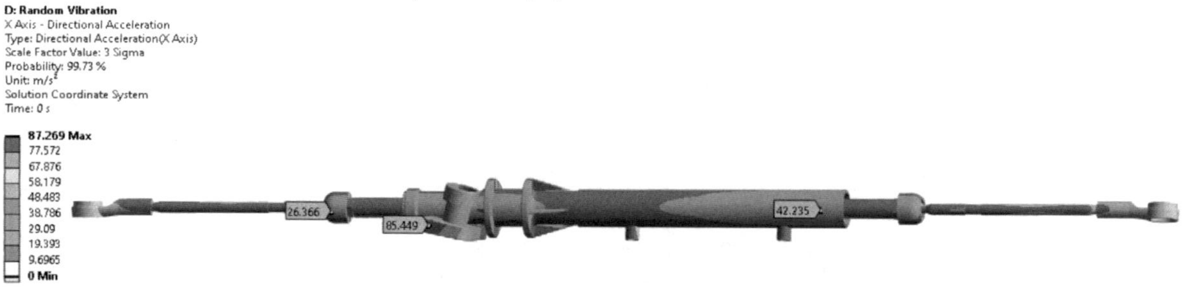

Figure 12. Random vibration simulation on X axis.

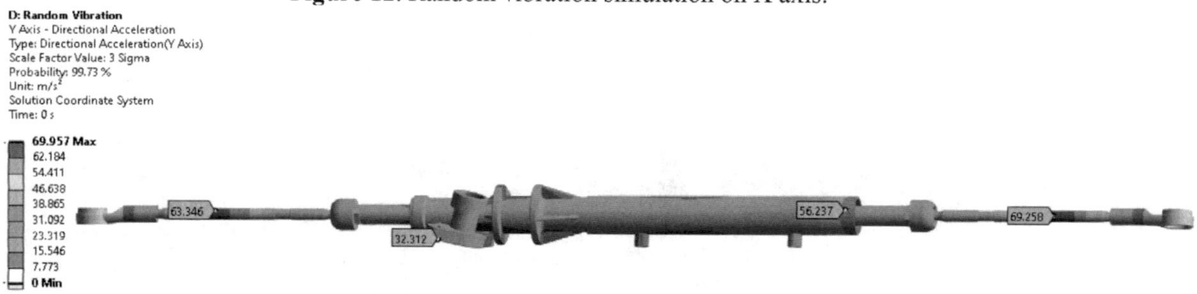

Figure 13. Random vibration simulation on Y axis.

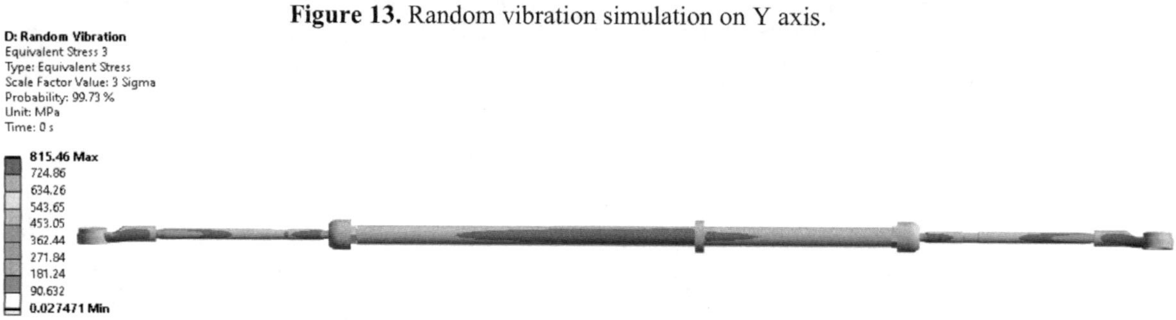

(a)

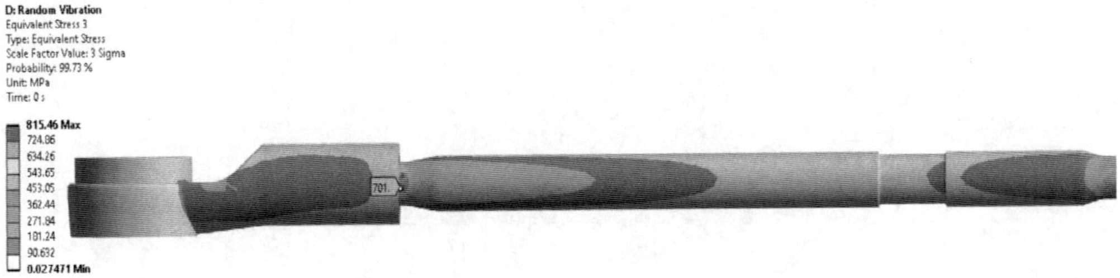

(b)

Figure 14. Random vibration simulation equivalent stress on the steering rack: a) Propagation on the whole steering rack; b) Most solicitated area.

4.2 Impact simulation

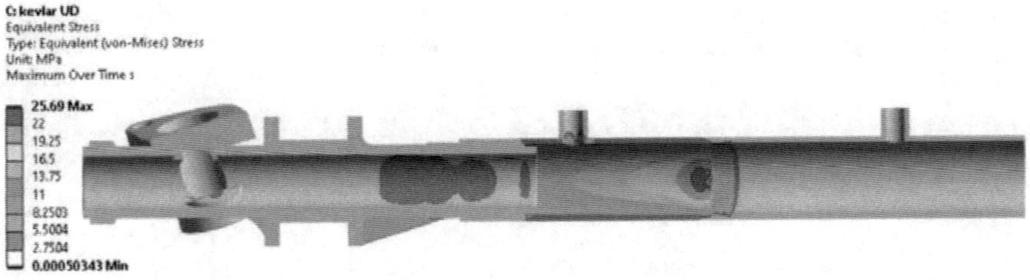

Figure 15. Equivalent stress of the steering housing.

5. High modulus carbon fiber

5.1 Random vibration simulation

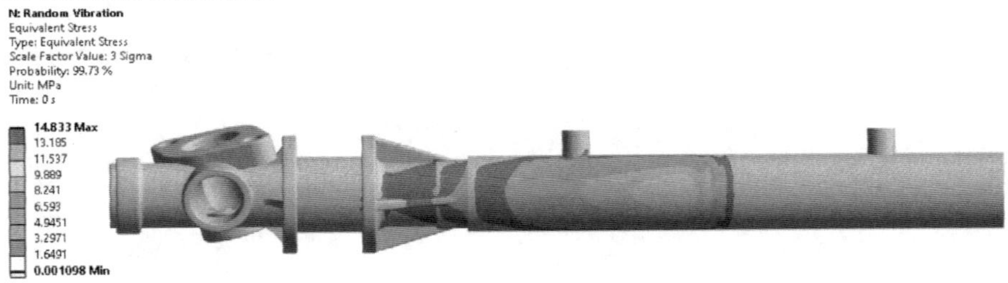

(a)

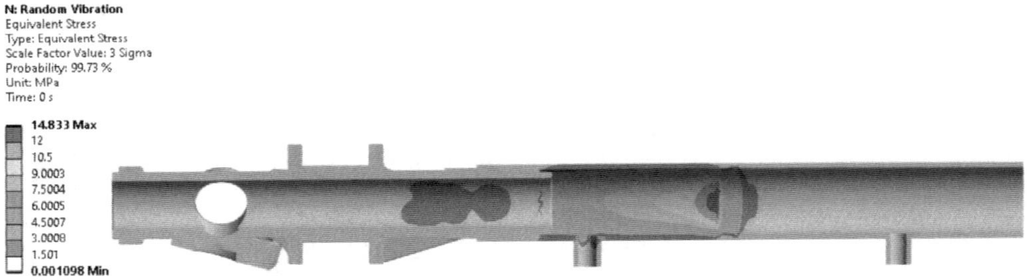

(b)

Figure 16. Random vibration simulation equivalent stress: a) Full view; b) Section view.

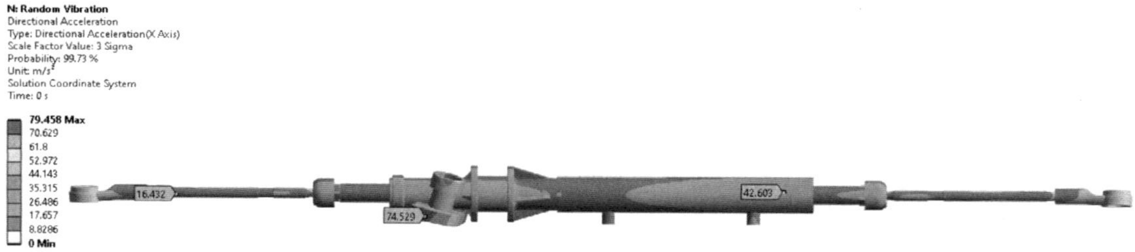

Figure 17. Random vibration simulation on X axis.

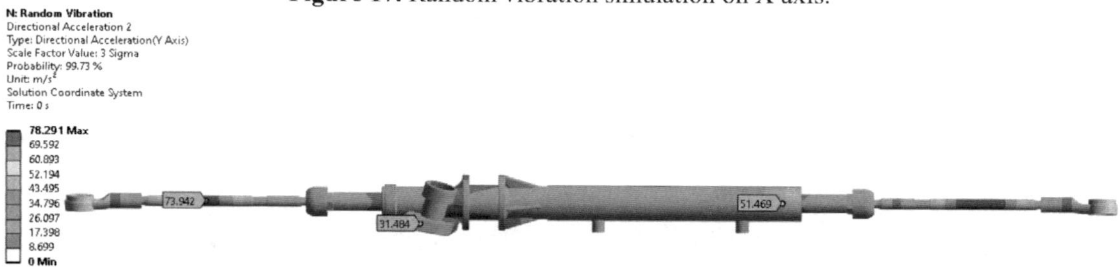

Figure 18. Random vibration simulation on Y axis.

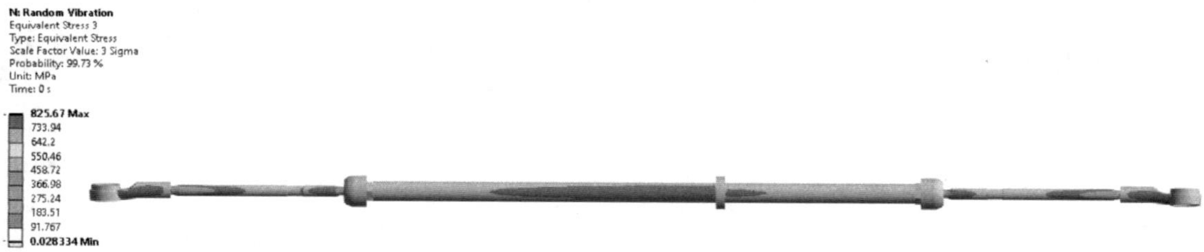

(a)

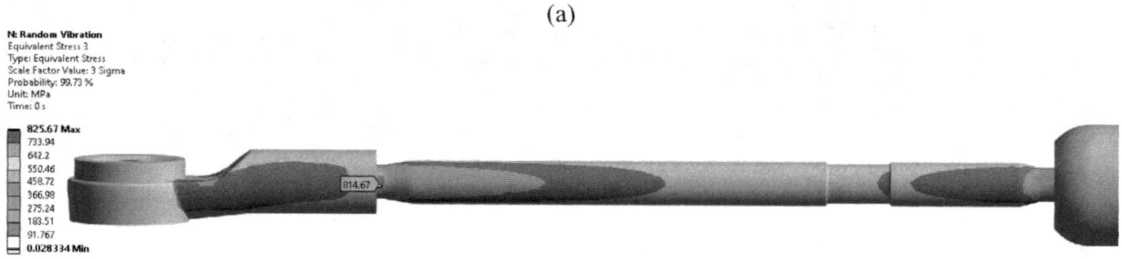

(b)

Figure 19. Random vibration simulation equivalent stress on the steering rack:
a) Propagation on the whole steering rack; b) Most solicitated area.

5.2 Impact simulation

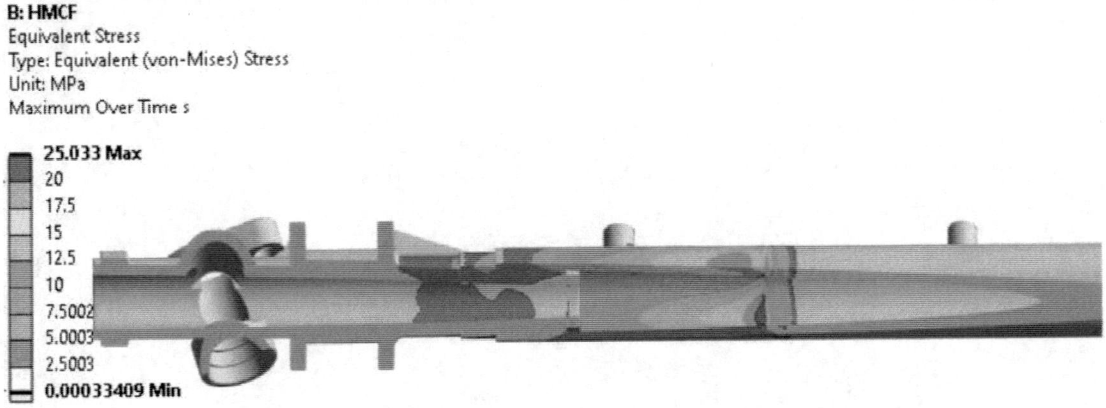

Figure 20. Equivalent stress of the steering housing.

6. Standard carbon fiber

6.1 Random vibration simulation

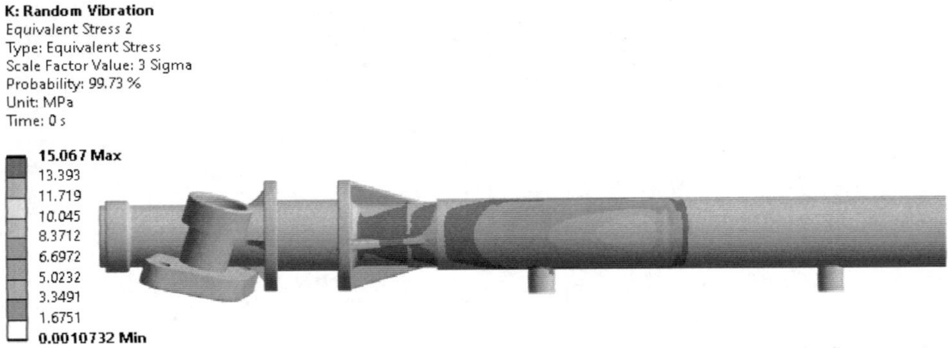

a)

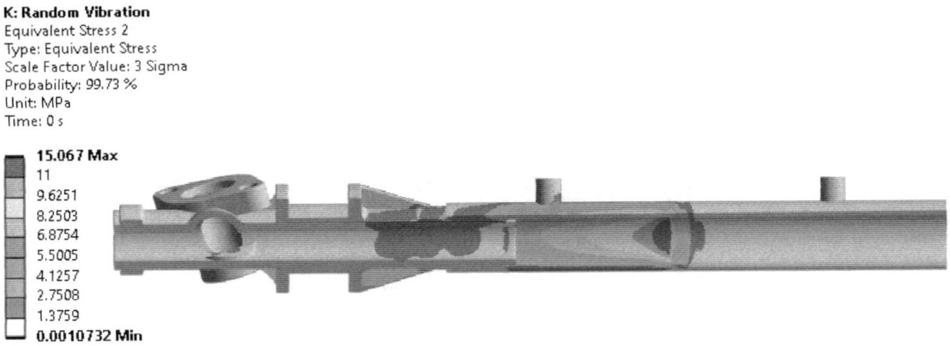

b)

Figure 21. Random vibration simulation equivalent stress:
a) Full view; b) Section view.

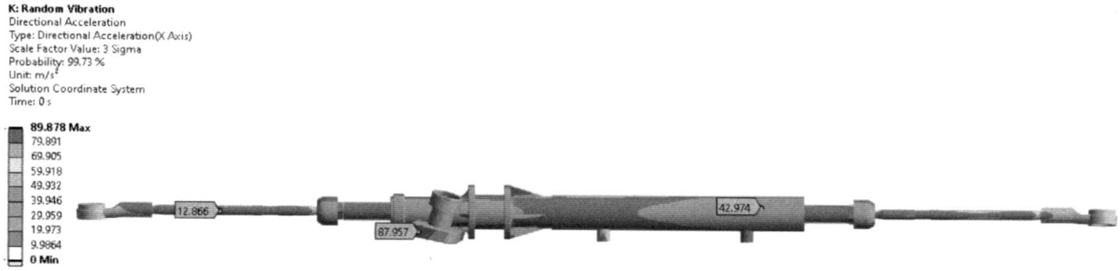

Figure 22. Random vibration simulation on X axis.

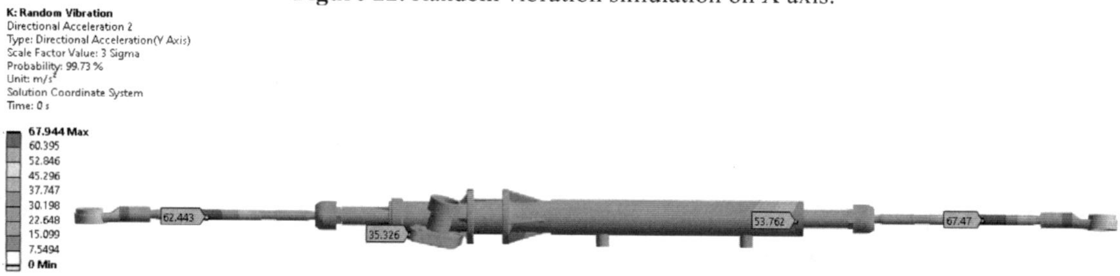

Figure 23. Random vibration simulation on Y axis.

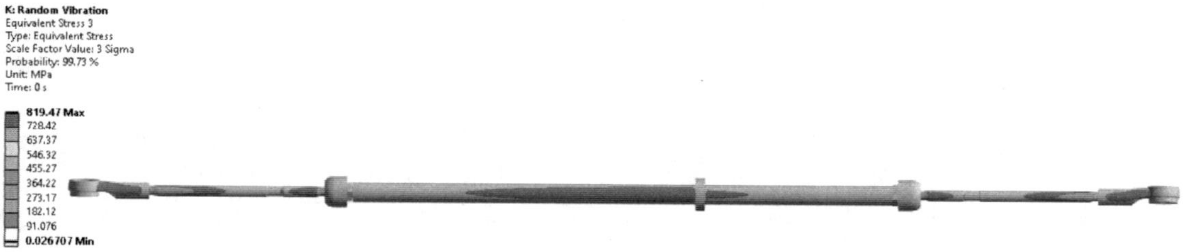

(a)

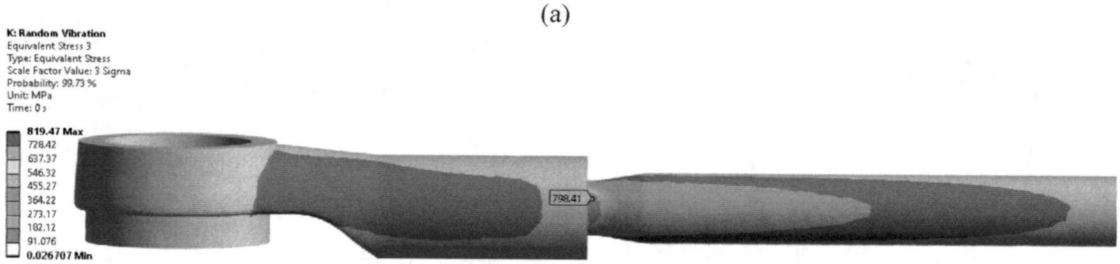

(b)

Figure 24. Random vibration simulation equivalent stress on the steering rack:
a) Propagation on the whole steering rack; b) Most solicitated area.

6.2 Impact simulation

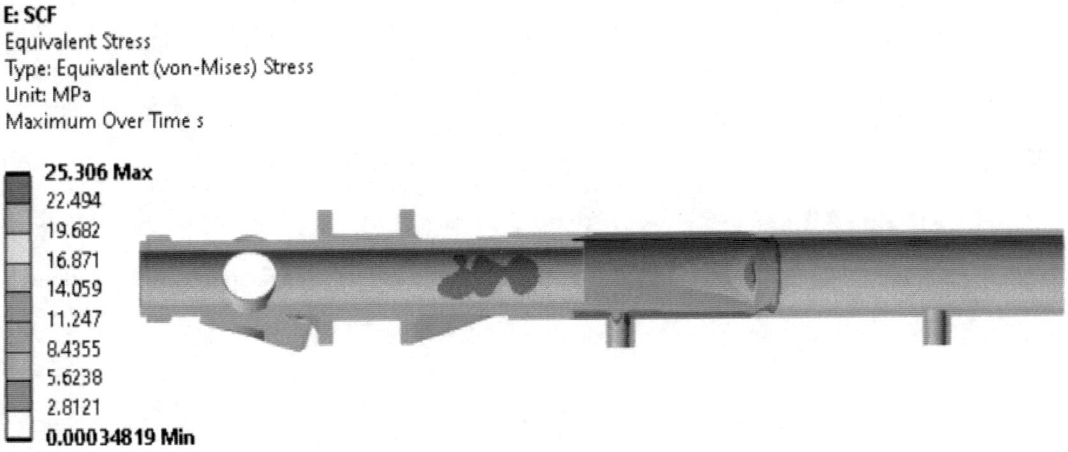

Figure 25. Equivalent stress of the steering housing.

7. Conclusion

Random vibration is a statistical analysis. Compared to an analysis that expresses tasks using a known force function and yields a valid result is not deterministic. Because there is no medium in the aleatory analysis of vibrations, the residual is zero. We used three standard deviations (3-sigma) in our simulations, which suggests that there is a 99.73% chance that the results would be smaller than or equal to the calculated value.

Based on the standard for Chassis random vibration tests for automobiles, the simulation was performed. It simulates the vibration that results in a steering system when an impact with a foreign object occurs.

The biggest value of the equivalent stress is found in the simulation where we used Al-Si8Cu3.5 at 23.03 MPa with High modulus carbon fiber having the smallest value at 14.83 MPa, showing that by simply replacing the steering rack housing material with Carbon fiber we can reduce the vibration by 35.6%.

For this simulation a hydraulic steering assisted rack was selected. Although the electric assisted steering is getting a lot of attention, the hydraulic assisted steering is more common in affordable cars.

We found that the vibration from the housing is transmitted through the piston all the way to the wheels. The most solicitated part of the whole assembly was the tie rod end.

Composite materials are lighter than metal alloys. By using composite materials in steering system construction, the weight of the vehicle is reduced, the fuel efficiency increases and as shown in this article, the vibration produced when an external stimulus is present can be reduced with up to 35.6% on the directly affected part and with up to 14,67% to the rest of the system.

During the impact simulation, the aluminum alloy, entered in the plastic deformation domain. This could cause the piston to fail and prevent the functioning of the assistance system or to the premature wear of the assembly. While the aluminum alloy was deformed, all composite materials were in the elastic deformation domain, maintaining their structural integrity.

While obtaining composite materials is a complex and costly process that is continuously improving, this type of material could prove revolutionary in the automotive industry. Due to their high resistance and improved elasticity and reduced weight this type of material could be used to replace common parts in the car to improve mileage and reduce pollution.

The simulations presented above showed similar values for both impact and random vibrations. Taking the price into consideration in comparison with the aluminum alloy (≈20 US Dollars per kg) normal

IOP Conf. Series: Materials Science and Engineering 1290 (2023) 012012 doi:10.1088/1757-899X/1290/1/012012

carbon fiber shows a slight advantage (30 US dollars per kg) compared to high modulus carbon fiber (38 US dollars per kg) and Kevlar (37.5 US Dollars per kg).

Even though high modulus carbon fiber showed better performance during impact simulation, carbon fiber should be used for this application due to its lower price and easier manufacturing technique.

Due to the rapid advancement of technology and the rise of concern for the environment, new technologies are being developed for use in the automotive industry. These technologies are being developed for some special applications like F1 cars, off-road and military vehicles. Also, the interest for the electric vehicle is on the rise. The main issue for this kind of vehicle is autonomy and life of the battery. Decreasing the weight of the vehicle could partially solve these issues. Right now, using composite materials for the common parts of the car is not feasible due to high costs and the complicated technological processes used to obtain this kind of material. But the number of research papers regarding composite materials is on the rise, and with each year the production costs and the costs of materials are decreasing, and in a few decades, we can assume that composite materials will be a part of our life.

8. References:

[1] Roylance D, Introduction to composite materials, Department of Materials Science and Engineering Massachusetts Institute of Technology Cambridge, March 24, 2000
[2] Harris B, Engineering Composite Materials, The Institute of Materials, London, 1999
[3] Reis P N B., Ferreira J A M, Santos P, Impact response of Kevlar composites with filled epoxy matrix
[4] Chae H G, Newcomb B A, Gulgunje P V, High strength and high modulus carbon fibers, 12 May 2015
[5] He Y, Tian G, Pan M, Chen D, Impact evaluation in carbon fiber reinforced plastic (CFRP) laminates using eddy current pulsed thermography, 4 November 2013
[6] Akangah P, Lingaiah S, Shivakumar K, Effect of Nylon-66 nano-fiber interleaving on impact damage resistance of epoxy/carbon fiber composite laminates, 13 November 2009
[7] Khan S U, Li C Y, Siddiqui N A, Kim J K, Vibration damping characteristics of carbon fiber-reinforced composites containing multi-walled carbon nanotubes, 23 June 2011
[8] Singha T J, Samantab S, Characterization of Kevlar Fiber and Its Composites: A Review, 4th International Conference on Materials Processing and Characterization
[9] ISO 10326-1:2016 - Mechanical vibration - Part 1: Basic requirements
[10] IEC 60068 - Environmental testing - Part 2-10: Tests - Test J and guidance: Mould growth

Comparative assessment of high efficiency flat solar thermal collectors' performances

Mihaela Ciocan[1], Alexandru Serban[1], Alexandru Dobrovicescu[1], Daniel P Hiris[2], Mircea C. Dudescu[2], Mugur C Balan[*,2]

[1]University Politehnica of Bucharest, Splaiul Independenţei nr. 313, Sector 6, 060042 Bucharest, Romania

[2]Technical University of Cluj-Napoca, Bd. Muncii 103-105, 400641, Cluj-Napoca, Romania

*mugur.balan@termo.utcluj.ro

Abstract. The study investigates the performances of high efficiency flat solar thermal collectors, under the climatic conditions of Galaţi, Romania. The same mathematical model was used to evaluate the stationary and the transitory thermal behavior of 12 high efficiency collectors, with reduced heat losses. The thermal characteristics of the collectors were extracted from the technical reports of two European independent testing laboratories. The annual specific energy production was found to vary in a large domain between 415.9 kWh/m^2 and 947.9 kWh/m^2, the maxim value representing more than the double of the minimum one. It was highlighted that both the optical efficiency and the heat losses contribute to the global efficiency of the flat solar thermal collectors, and the most efficient one is the flat evacuated one.

Key words: Solar thermal collectors; Mathematical modelling; Efficiency; Heat losses; Energy production.

1. Introduction

The use of solar energy in general, and particularly the solar thermal energy, became a very important issue in the context of the actual complex energy context, dominated by the problem of the energy security, but also by the global warming, reduction of the CO_2 emissions and increasing of the energy efficiency.

The solar thermal collectors (STC) are equipment dedicated to the solar energy conversion into heat embedded in a thermal agent [1]. There are several types of STC [2], and their efficiency depends by the capacity of the component materials to absorb the incident solar radiation, and by the thermal losses into the environment [3]. Between the different technologies, the flat solar thermal collectors (FSTC) are the simplest, but also enough efficient to ensure a high popularity on the market, because of their capability to provide hot water up to (100 – 120) °C or even more [4].

Some solutions to increase the efficiency of FSTC are concerning the absorber surface that can be improved by multiple-passes of air, by the use of porous materials, by using air-jet impingement by integration of heat storage materials, etc. [5]. Future trends related to the FSTC will focus on

Content from this work may be used under the terms of the Creative Commons Attribution 3.0 licence. Any further distribution of this work must maintain attribution to the author(s) and the title of the work, journal citation and DOI.
Published under licence by IOP Publishing Ltd

integration of new materials with different constraints: cost effectiveness, recyclability, environmental impact, etc. [6]. The use of nanofluids can also improve the performances of FSTC [7]. The effect of the types and number of covers was investigated in [8]. Passive performers enhancers and the influence of magnetic field were studied in [9] and [10] respectively. The importance of the FSTC integrated in buildings is raised in [11–14].

Experimental studies of FSTC are presented in [15] or in [16] while mathematical modelling of the thermal behavior of different types of FSTC is approached in [1,3,4,17]. A new technology of evacuated FSTC is investigated in [18] and [19].

The goal of this study is to investigate a comparative analysis of the performances of 12 models of FSTC of which one is evacuated, in the climatic conditions of Galați, Romania. To the best of the authors' knowledge, such a study is new and is missing from the literature. The study continues the previous author's work [20] or [21].

2. Material and method

Different highly efficient FSTC were selected for the comparison, based on the testing data sheets available on the web sites of two independent testing laboratories: Solartechnik Prüfung Forschung of Swiss (https://www.spftesting.info/data/1.kollektoren/) and Solar Keymark of Netherlands (http://www.solarkeymark.nl/DBF/#).

The climatic data of Galați, Romania, situated at the coordinates 45.4353° N, 28.0080° E, were extracted from the web site of the European Union (https://re.jrc.ec.europa.eu/pvg_tools/en/tools.html). It was used the typical meteorological year (TMY): PVGIS-SARAH2: 2005 – 2020.

The global solar radiation on the horizontal plane and the ambient temperature, for the location of Galați, Romania, are presented in figure 1.

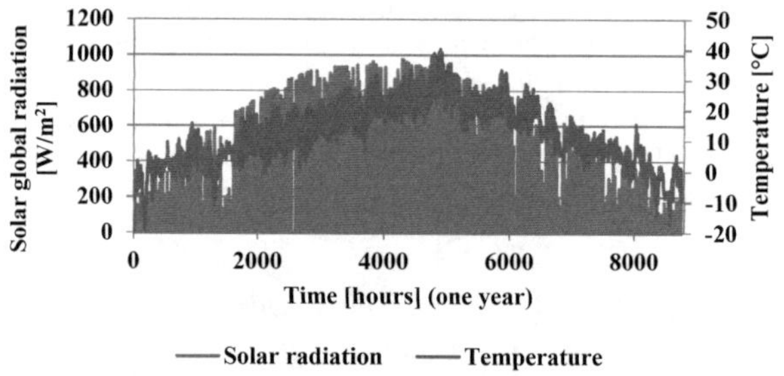

Figure 1. The global solar radiation on the horizontal plane and the ambient temperature for Galați, Romania

The efficiency (η [-]) of the FSTC can be computed as [4,17,21]:

$$\eta = \eta_0 - k_1 \frac{\Delta t}{I_{gt}} - k_2 \frac{\Delta t^2}{I_{gt}} \tag{1}$$

where η_0 [-] is the optical efficiency of the FSTC; k_1 [W/m²K] and k_2 [W/m²K²] are correction factors depending on the thermal losses; Δt [°C] is the difference between the average temperature of the heat transfer fluid inside the FSTC and the ambient temperature, and I_{gt} [W/m²] is the incident solar radiation, normal on the plane of the FSTC. The incident solar radiation, normal on the plane of the FSTC, I_{gt} [W/m²], was calculated based on the algorithm presented in [4,17,21], considering that the FSTC are oriented facing South with a tilt angle of 35 °.

IOP Conf. Series: Materials Science and Engineering 1290 (2023) 012013 doi:10.1088/1757-899X/1290/1/012013

The values of the characteristic constructive parameters for the selected FSTC, in random order, including the aperture (A_0 [m^2]), are presented in table 1.

Table 1. Aperture, optic efficiency, and correction factors for the considered FSTC

Code	Collector type	Collector model	Country	Correction coefficients			
				A_0 [m^2]	η_0 [%]	k_1 [Wm^{-2}K^{-1}]	k_2 [Wm^{-2}K^{-2}]
C01	Flat plate	AMX AR 2.85	POL	2.64	0.772	2.22	0.0402
C02	Flat plate	SF500-15DG-M	FIN	14.83	0.861	2.51	0.0086
C03	Flat plate	FP-3	CHN	2.84	0.771	1.77	0.0539
C04	Flat plate	Megaslate II	CHE	0.78	0.694	1.89	0.0441
C05	Flat plate	FINO	DEU	0.90	0.707	2.25	0.0241
C06	Flat plate	Fyrosol KA 1	CHE	2.36	0.774	2.98	0.0156
C07	Flat plate	UHV Collector C0	CAN	1.80	0.746	1.34	0.0002
C08	Flat plate	TIGI HC1-A	ISR	1.80	0.785	2.20	0.0036
C09	Flat plate	SF500-15	FIN	14.83	0.874	3.16	0.0098
C10	Flat plate	Integra	CHE	2.03	0.807	2.74	0.0190
C11	Flat plate	HT-SolarBoost 35/10	DNK	12.56	0.838	2.46	0.0197
C12	Flat evacuated	MT-POWER v4	CHE	1.84	0.780	0.53	0.0064

The codes provided in the first column table are used in the study to further identify the FSTC models. All the selected FSTC are of high efficiency, due to the low values of k_1 (lower than 3.2 W/m^{2}K), in the context in which most of the FSTC available on the market presents values of k_1 higher than 4 W/m^{2}K.

3. Results and discussion

The efficiency of the FSTC as function of the temperature difference (Δt [°C]), between the average temperature of the heat transfer fluid inside the FSTC and the ambient temperature, for the incident solar radiation, normal on the plane of the FSTC, I_{gt} = 750 W/m^2, for all the considered FSTC, is presented in figure 2.

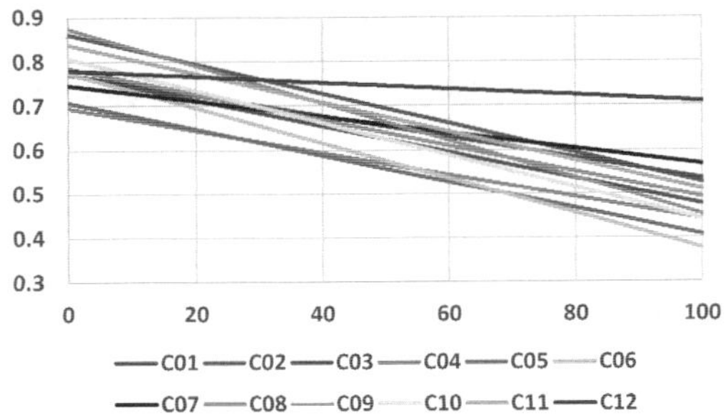

Figure 2. The efficiency of the selected FSTC

The following global characteristics of the investigated FSTC can be extracted:

- The low values of k_2, determines the shape almost linear, instead of parabolic, as it was expected, of the efficiency variation, for almost all the FSTC.
- C02, C09 and C11 presents the higher optical efficiencies, and thus, the global efficiency starts from the higher values.
- C02, C06 and C09 presents the higher values of k_1 corresponding to the higher thermal losses and the efficiency is decreasing with the higher speed.

IOP Conf. Series: Materials Science and Engineering 1290 (2023) 012013 doi:10.1088/1757-899X/1290/1/012013

- C07 and C12 presents the lower values of k_1 corresponding to the lower thermal losses and the efficiency is decreasing with the lower speed, the characteristic being closer to the horizontal, than for the rest of the FSTC.
- C12 presents the closest to the constant efficiency, because $k_1 = 0.53$ W/m^{2}K^2, much lower than the values of k_1 for the other FSTC, situated in the range (1.34 – 2.98) W/m^{2}K^2.

The technology of evacuated FSTC, present only at the collector C12 determines the higher values of the efficiency, for $\Delta t > 30$ °C.

The hourly variation of the efficiencies, of the considered SFTC, for the whole year, under the real climatic conditions of Galați, are presented for groups of 3 FSTC, in figures 3 – 6.

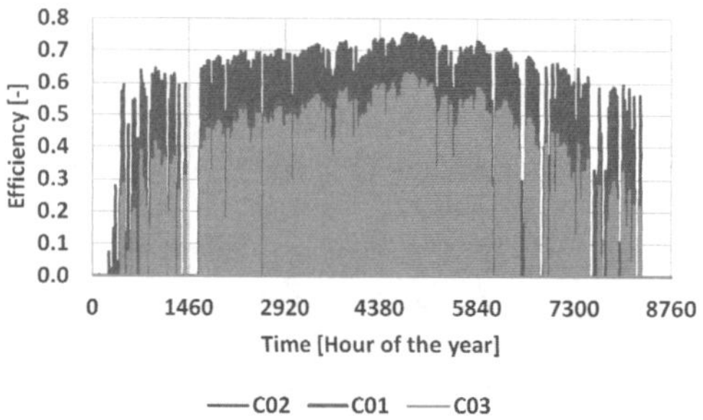

Figure 3. The hourly variation of the efficiency for C01-C03

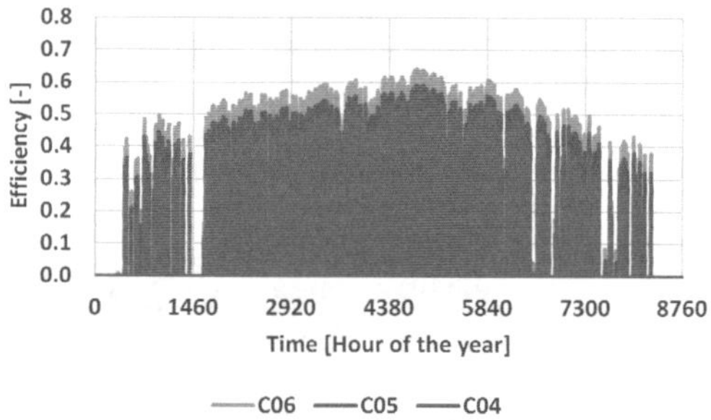

Figure 4. The hourly variation of the efficiency for C04-C06

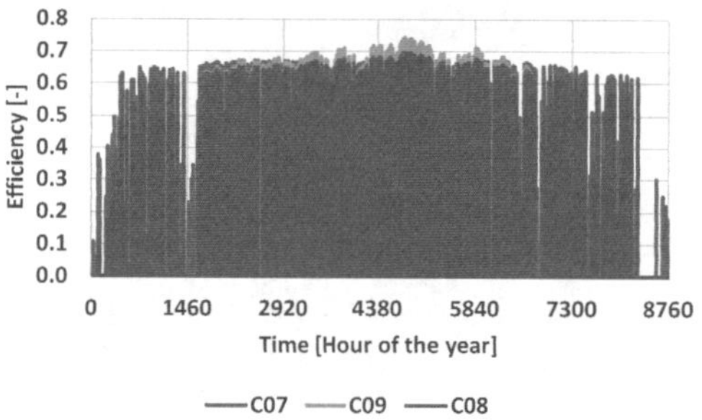

Figure 5. The hourly variation of the efficiency for C07-C09

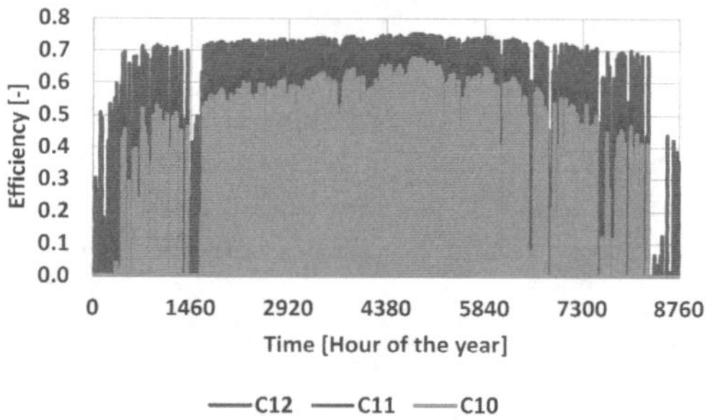

Figure 6. The hourly variation of the efficiency for C10-C12

The order of the FSTC on the charts and the colors (identical on the charts), were selected in such a way to be visible the efficiency variation of all the FSTC. It can be observed that some of the FSTC, such as C07 or C12, presents a lower efficiency, while other SFTC, such as C03 or C04, presents high efficiency. The maximum efficiencies range between 56.2 % for C04 and 75.4 % for C02.

To compare the heat production of the considered FSTC, it was analyzed the monthly and annual specific energy production, instead of the total energy productions, because the size of the collectors is very different, since the aperture (A_0 [m^2]) range between 0.78 m^2 for C04 and 14.83 m^2 for C02 and C09.

The monthly specific energy production of the FSTC for the considered collectors, in the order of the annual heat production is presented in figure 7.

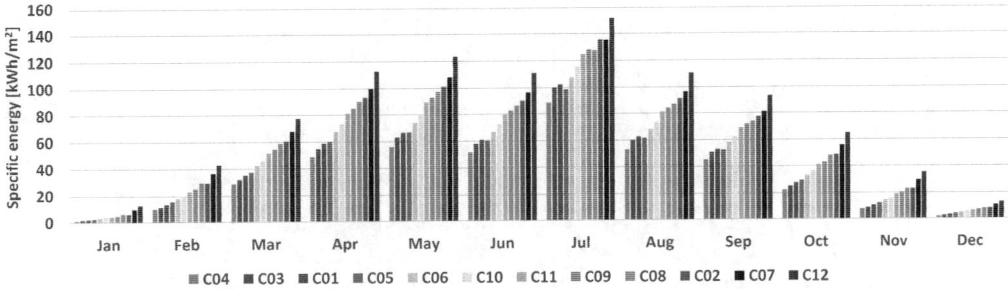

Figure 7. The monthly specific energy production for the FSTC

Even if all the considered FSTC are of high efficiency, due to the reduced heat losses, there are important differences, roughly up to double, between the monthly energy productions.

Figure 8 presents the annual specific energy production of the considered FSTC.

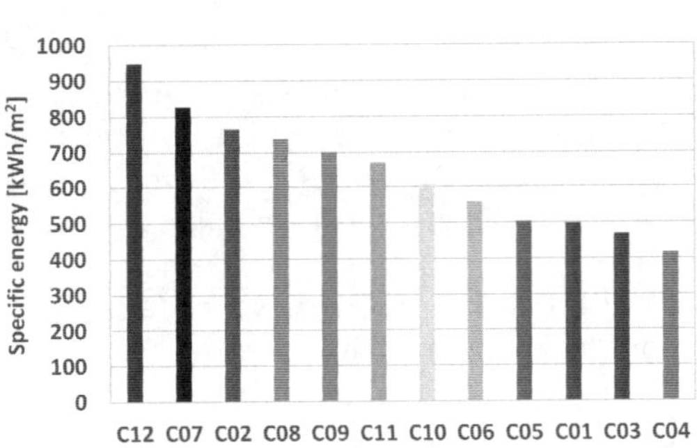

Figure 8. The annual specific energy production of the FSTC

The lower yearly specific energy production of 415.9 kWh/m^2 corresponds to C04, and the higher yearly specific energy production of 947.9 kWh/m^2 corresponds to C12 and is more than double the production of C04. The high and unexpected range of the yearly specific energy production, between these highly efficient FSTC, impose to the investigation of the causes leading to this situation. Annual specific energy production higher than 600 kWh/m^2 is provided by 7 FSTC while the rest FSTC produces less, but only 4 FSTC produces more than 700 kWh/m^2 and only 2 FSTC produces more than 800 kWh/m^2.

According to (eq. 1), the most important parameters on which depends the efficiency are the constructive ones: the optical efficiency η_0 [-] and the correction factor k_1 [W/m^{2}K], but also the operating parameters: the temperature difference Δt [°C] and the incident solar radiation, normal to the plane of the FSTC I_{gt} [W/m^2]. The operating parameters depends by the location, but on each location, they have similar effect for all the FSTC, thus the efficiency and the specific energy production of the FSTC in a particular depends mainly by the constructive parameters η_0 and k_1.

The optical efficiency and the correction factor k_1 of the considered SFTC are presented in figures 9 and 10 respectively.

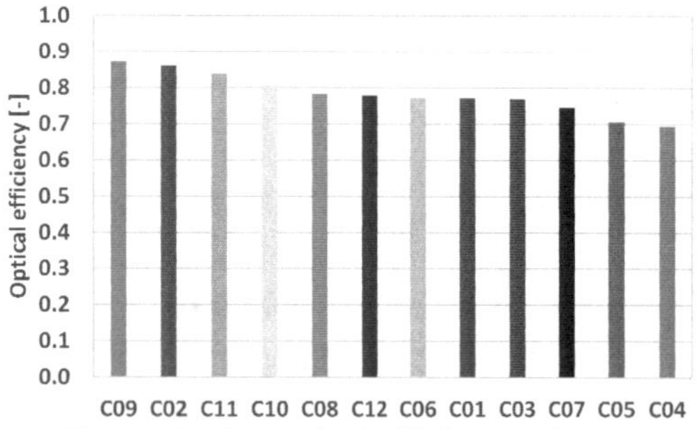

Figure 9. The optical efficiency of the FSTC

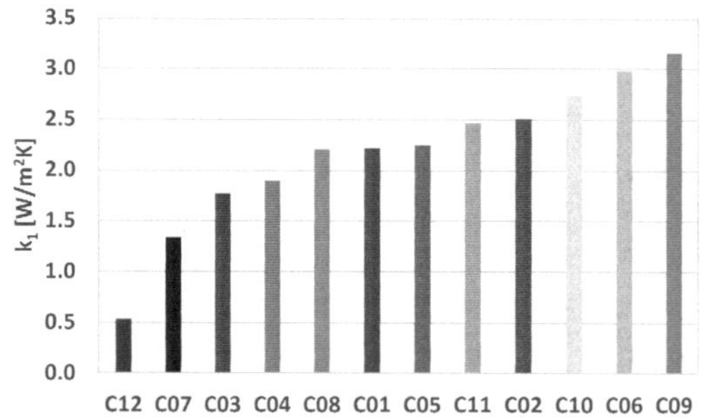

Figure 10. Correction factor k_1 of the FSTC

The optical efficiency range between 69.5 % for C04 and 87.4 % for C10, the absolute difference being 18 % representing 20.6 % of the maximum optical efficiency. The correction factor k_1 range between 0.53 W/m²K for C12 and 3.16 W/m²K for C09, the absolute difference being 2.63 W/m²K, representing 83.2 % of the higher k_1.

Considering the values of these two parameters, the correction factor k_1 seems to be more important than the optical efficiency, because the variation range is higher. This hypothesis is confirmed by the FSTC C12 and C07, having the lower values of k_1 and the higher annual specific energy production even if C07 have one of the lower optical efficiencies. On the other side, FSTC C09 and C06, having the higher values of k_1 are in the middle range of the annual specific energy production, not in the lower one as it should be expected.

The importance of the optical efficiency is suggested by C02, having the second higher optical efficiency and a high correction factor k_1 (high heat losses) but it is the third FSTC by the annual specific energy production. C09 having the higher optical efficiency but also the higher k_1 (the higher heat losses), is the fifth FSTC by the annual specific energy production, also proving the important effect of the optical efficiency. In the group of the FSTC with comparable optical efficiencies C08, C12, C06, C01 and C03, the order of the correction factors k_1 is C12, C03, C08, C01 and C06 (from lower to higher), but the order of the annual specific energy production is C12, C08, C06, C01 and C03 (from higher to lower), meaning that the operating conditions (even if similar), can favorize one FSTC or another one, by situating them for longer or shorter periods of time in the most favorable operating conditions.

4. Conclusion

The study presents a comparative investigation, under the climatic conditions of Galati, Romania, of the performances of 12 models of FSTC with high energy efficiency determined by low heat losses. The constructive characteristic parameters of these equipment were extracted from the available testing data sheets available form two European independent testing laboratories.

The yearly specific energy production range is (415.9 - 947.9) kWh/m² being higher than 600 kWh/m² for 7 FSTC. Only 4 FSTC produces more than 700 kWh/m² and only 2 FSTC produces more than 800 kWh/m².

The higher annual specific energy production corresponds to the FSTC C12, the single one with yearly specific energy production higher than 900 kWh/m² and the single one being evacuated. The higher efficiency of this FSTC proves the superiority of this technology.

The first two higher annual specific energy production, corresponds to the collectors with the lower heat losses (C12 and C07), but the collectors with the higher optical efficiencies (C09 and C02) even if they have important heat losses (C09 having even the higher heat losses), are providing the 3ed (C02) and 5th (C09) annual specific energy production, highlighting that the optical efficiency have at less the same importance than the heat losses for the global efficiency of the FSTC.

It was proved that the simple analysis of the characteristic parameters of the FSTC is not sufficient to estimate the best performances in a particular application. The local climatic conditions may present an important influence and only a detailed investigation, eventually doubled by an economic analysis, allow the best choice for a particular FSTC to be used in a particular application. The investigation performed in this study represents an example of the way in which a detailed technical investigation can be conducted to compare the performances of several FSTC under the climatic data of a particular location.

Acknowledgement

This study was partially financially supported by the Project "Network of excellence in applied research and innovation for doctoral and postdoctoral programs / InoHubDoc", project co-funded by the European Social Fund financing agreement no. POCU/993/6/13/153437.

This study was partially supported by a grant of the Ministry of Research, Innovation and Digitization, CCCDI - UEFISCDI, project number PN-III-P2-2.1-PED-2021-0544, within PNCDI III (PED 706).

References

[1] Kalogirou SA. Solar Energy Engineering: Processes and Systems: Second Edition. Solar Energy Engineering: Processes and Systems: Second Edition 2014:1–819. https://doi.org/10.1016/C2011-0-07038-2.

[2] Evangelisti L, De Lieto Vollaro R, Asdrubali F. Latest advances on solar thermal collectors: A comprehensive review. Renewable and Sustainable Energy Reviews 2019;114. https://doi.org/10.1016/j.rser.2019.109318.

[3] Goswami DY. Principles of Solar Engineering. Principles of Solar Engineering 2015. https://doi.org/10.1201/B18119.

[4] Duffie J, Beckman W. Solar engineering of thermal processes. 1980.

[5] Vengadesan E, Senthil R. A review on recent developments in thermal performance enhancement methods of flat plate solar air collector. Renewable and Sustainable Energy Reviews 2020;134. https://doi.org/10.1016/j.rser.2020.110315.

[6] Colangelo G, Favale E, Miglietta P, De Risi A. Innovation in flat solar thermal collectors: A review of the last ten years experimental results. Renewable and Sustainable Energy Reviews 2016;57:1141–59. https://doi.org/10.1016/j.rser.2015.12.142.

[7] Zayed ME, Zhao J, Du Y, Kabeel AE, Shalaby SM. Factors affecting the thermal performance of the flat plate solar collector using nanofluids: A review. Solar Energy 2019;182:382–96. https://doi.org/10.1016/J.SOLENER.2019.02.054.

[8] Dagdougui H, Ouammi A, Robba M, Sacile R. Thermal analysis and performance optimization of a solar water heater flat plate collector: Application to Tétouan (Morocco). Renewable and Sustainable Energy Reviews 2011;15:630–8. https://doi.org/10.1016/J.RSER.2010.09.010.

[9] Murugan M, Saravanan A, Elumalai P V., Kumar P, Ahamed Saleel C, Samuel OD, et al. An overview on energy and exergy analysis of solar thermal collectors with passive performance enhancers. Alexandria Engineering Journal 2022;61:8123–47. https://doi.org/10.1016/J.AEJ.2022.01.052.

[10] Shojaeizadeh E, Veysi F, Zareinia K, Mansouri AM. Thermal efficiency of a ferrofluid-based flat-plate solar collector under the effect of non-uniform magnetic field. Appl Therm Eng 2022;201. https://doi.org/10.1016/J.APPLTHERMALENG.2021.117726.

[11] Kayali H, Alibaba AsstPDrH. Comparison of Different Solar Thermal Energy Collectors and Their Integration Possibilities in Architecture. Sustainability in Environment 2017;2:36. https://doi.org/10.22158/SE.V2N1P36.

[12] Unguresan P V., Porumb RA, Petreus D, Pocola AG, Pop OG, Balan MC. Orientation of Facades for Active Solar Energy Applications in Different Climatic Conditions. Journal of Energy Engineering 2017;143:04017059. https://doi.org/10.1061/(ASCE)EY.1943-7897.0000486.
[13] Visa I, Moldovan M, Duta A. Novel triangle flat plate solar thermal collector for facades integration. Renew Energy 2019;143:252–62. https://doi.org/10.1016/j.renene.2019.05.021.
[14] Visa I, Moldovan M, Comsit M, Neagoe M, Duta A. Facades Integrated Solar-thermal Collectors - Challenges and Solutions. Energy Procedia 2017;112:176–85. https://doi.org/10.1016/J.EGYPRO.2017.03.1080.
[15] Salavati Meibodi S, Kianifar A, Niazmand H, Mahian O, Wongwises S. Experimental investigation on the thermal efficiency and performance characteristics of a flat plate solar collector using SiO2/EG–water nanofluids. International Communications in Heat and Mass Transfer 2015;Complete:71–5. https://doi.org/10.1016/J.ICHEATMASSTRANSFER.2015.02.011.
[16] Farajzadeh E, Movahed S, Hosseini R. Experimental and numerical investigations on the effect of Al2O3/TiO2[sbnd]H2O nanofluids on thermal efficiency of the flat plate solar collector. Renew Energy 2018;118:122–30. https://doi.org/10.1016/j.renene.2017.10.102.
[17] Quaschning V. Understanding Renewable Energy Systems. 2007.
[18] Moss R, Shire S, Henshall P, Arya F, Eames P, Hyde T. Performance of evacuated flat plate solar thermal collectors. Thermal Science and Engineering Progress 2018;8:296–306. https://doi.org/10.1016/J.TSEP.2018.09.003.
[19] Caldarelli A, Gaudino E, De Luca D, Farooq U, Musto M, Di Gennaro E, et al. Low emissivity thin film coating to enhance the thermal conversion efficiency of selective solar absorber in high vacuum flat plate collectors. Thin Solid Films 2023;764. https://doi.org/10.1016/J.TSF.2022.139632.
[20] Bălan MC, Jäntschi L, Bolboacă S, Damian M. Thermal Solar Collector Behaviour in Romania. Pol J Environ Stud 2010;19:231–41.
[21] Pop OG, Magurean AM, Pocola AG, Ciocan M, Zaaoumi A, Balan MC. Performances of Solar Thermal Collectors in Different Climatic Conditions. Springer Tracts in Civil Engineering 2021:235–73. https://doi.org/10.1007/978-3-030-57418-5_9.

IOP Conf. Series: Materials Science and Engineering 1290 (2023) 012014 doi:10.1088/1757-899X/1290/1/012014

CO_2 energy vector in the concept of circular energy

L Mihaescu[3], G Lazaroiu[1], U Desideri[2], G Negreanu[3], I Pisa[3] and R M Grigoriu[1,4]

[1] University "Politehnica" of Bucharest, Department of Energy Generation and Use, 313 Splaiul Independenței, 060042 Bucharest, Romania

[2] University of Pisa, Department of Thermal Machines, Lungamo Pacinotti 43, 56126 Pisa, Italy

[3] University "Politehnica" of Bucharest, Department of Technical Thermodynamics, Engines, Thermal and Refrigeration Equipment, 313 Splaiul Independenței, 060042 Bucharest, Romania

[4] Author to whom any correspondence should be addressed

Email: manuela.grigoriu.1968@gmail.com

Abstract. The paper analyzes the conversion of the carbon dioxide (CO_2) through recovery and storage into an energy production vector by transforming it into methanol (CH_3OH) and carbon monoxide (CO). These gaseous components can produce energy by combustion, and the resulting carbon dioxide is reintroduced into the capture and storage circuit. This type of carbon dioxide transformation into an energy vector in a circular energy system will contribute to the elimination or to the significant reduction of its long-term storage in caves or under the sea water. By eliminating storage, the costs will be significantly reduced, as well as the environmental pollution. The paper considers the methanol use in residential heating instead of liquid hydrocarbons, especially for isolated areas or without a natural gas network, and also in the internal combustion engines in transport, agriculture and construction. The first direction of the research included an experimental phase that involved using a 55kW boiler for medium-sized residential heating. The second direction analyzed the transformation of CO_2 into CO (fuel gas) through gasification technology. For this transformation, the fixed bed gasification process was considered. The carbon monoxide can be used in a pure state or mixed with methane or other combustible gases, both in power plants and in internal combustion engines. The current significant results in the methanol production field enables CO_2 to become an energy vector, instead of a gas with zero potential and expensive to store. Thus, the research can significantly contribute to the development of power engineering as a circular application in the context of circular economy.

1. General considerations

The transformation of carbon dioxide (CO_2) into an energy vector involves its transformation into methanol (CH_3OH) or carbon monoxide (CO) and their use for energy production. Circular energy requires the following developments for this model: reducing CO_2 emissions during energy production by recirculating this component, reduction of stored CO_2 amounts (after its capture).

Figure 1 shows the scheme of the recirculated CO_2 circuit within the concept of using it as an energy vector.

Content from this work may be used under the terms of the Creative Commons Attribution 3.0 licence. Any further distribution of this work must maintain attribution to the author(s) and the title of the work, journal citation and DOI.
Published under licence by IOP Publishing Ltd

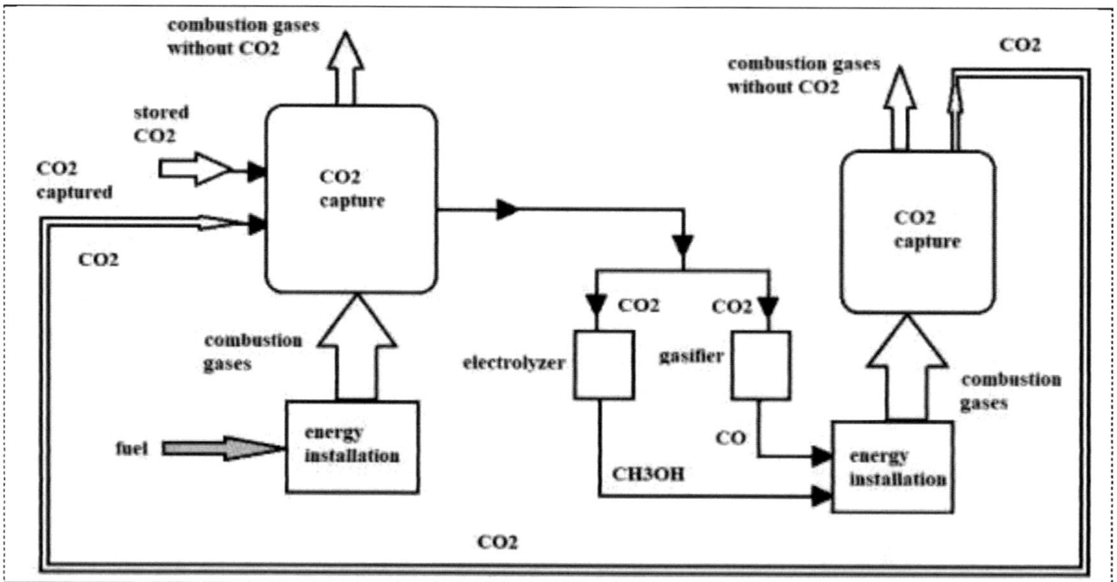

Figure 1. The concept of using carbon dioxide as an energy vector within a circular energy system.

At the current level of industrial development, methanol has greatly increased its applicability, surpassing previous traditional fields. This is mainly due to the achievements in its production field, but also due to the widening of its applicability field. With regard to energy applications, the realization of methanol fuel cells worths mentioning, alongside with direct combustion, so that over 45% of methanol production, which already shows an annual increase of about 8%, is intended for energy.

A significant achievement in the methanol use for energy applications is represented by the Solid Oxid Electrochemical Cell (SOEC) production technology, which has a great efficiency. The possible production of methanol from renewable sources completes the spectrum of its ecological use. As in the case of hydrogen, currently the green methanol notion was also introduced [1], [2].

Gasification is also a solution for methanol production. Gasification technology is also considered for the production of CO, or, by adaptation, for the conversion of CO_2 to CO. The technology for obtaining methanol (CH_3OH) from CO_2 is relatively new (relation 1), and it tends to replace a less ecological technology that uses carbon monoxide hydrogenation [1], [2].

$$\begin{cases} CO_2 + 3H_2 = CH_3 + H_2O \\ H_2 + CO_2 = CO + H_2O \end{cases} \tag{1}$$

By using the CuO–ZnO–Al_2O_3 catalyst at a pressure of 78 bar and a temperature of 290°C approximately 21% conversion degree is achieved in a single pass. In the end, a mass proportion of methanol of 63% and water of 37% is achieved (water separation does not represent a problem).

In order to converge towards the CO_2 energy vector notion, the paper is further analyzing the possibility of using methane in residential heating, to replace liquid hydrocarbons, alongside with the technology of converting CO_2 into CO through gasification in a fixed layer [2], [3], [4]. The researchers from Politehnica University of Bucharest have experience in these two fields, also having available experimental stands for studying different applications.

Another application related to methanol is represented by its combustion in internal combustion engines, with application in transport and agriculture, but also for the production of energy with their help. Generators for electricity production can be operated with methanol instead of liquid hydrocarbons, a fuel that incorporated CO_2 during its production process, thus contributing to carbon neutralization [5], [6], [7].

2. Analysis of the possibility of using methanol in residential heating installations

The analysis refers to the use of methanol in installations with reduced thermal output up to 1MW (Pt < 1MW), intended for residential heating. Wall-mounted heating plants were not taken into account, but those with thermal powers greater than 30 kW, intended for heating spaces of over 250 m^2 and up to 1000 kW since their necessary light liquid fuel no longer satisfies the environmental conditions, especially regarding the CO_2 emission.

Previously, experiments were carried out on a 55 kW boiler in order to analyze the combustion of non-conventional liquid fuels such as pure vegetable oils, a mixture of vegetable oils with light hydrocarbons and animal fats. The experience gained will be capitalized by performing ~~on~~ a suitable combustion of methanol [8], [9], [10].

Figure 2 shows the experimental installation that includes the 55 kW pilot boiler for residential heating. The boiler is equipped with a GB-Ganz type burner, return flow regulation system and electric heating of the fuel up to 75-80°C.

Figure 2. The 55kW pilot boiler for residential heating.

It is worth mentioning that methanol has a lower calorific value than liquid hydrocarbons – 22.7 MJ/kg compared to about 40MJ/kg. For a certain energy installation, this difference of about 45% will require an increased methanol flow in order to obtain the same thermal effect.

The main problem in using methanol is the possibility of using liquid hydrocarbon burners. A complete study will have to answer the following questions [8]:

- the spraying quality for the pressure in the liquid hydrocarbon burners from the residential heating;
- the methanol preheating influence on the viscosity variation and the determination of methanol compatibility for light hydrocarbons burners, which have a 70-80°C preheating;
- ignition capacity;
- the agreement between the combustion speed and the air speed for the current burners;
- the temperature field in the flame;
- variation in the residential heating installations efficiency when changing the liquid fuel;
- pollutant emissions.

As a first stage of study, the Fluent numerical simulation program can be used. The second stage can be achieved by using pilot installations, of certain powers, intended for residential heating, such as the one shown in Figure 2 [7].

IOP Conf. Series: Materials Science and Engineering 1290 (2023) 012014 doi:10.1088/1757-899X/1290/1/012014

3. Conversion of CO_2 to CO through gasification for energy production

The research in this section aimed to CO_2 conversion into CO by using the gasification technology. This installation is completely independent from the installation in section 2. They both have the same purpose, namely integrating (transforming) the carbon dioxide into a burnable fuel.

Carbon dioxide can be chemically reduced by an apparently simple solution, by the reaction in the carbon existing in the form of a heated layer (the principle in gasification plants) by the reaction [11], [12], [13]:

$$CO_2 + C = 2CO \text{ (endothermic reaction)} \tag{2}$$

The complexity of the solution lies in the maintenance of the ember layer during operation. If the start is made by the oxidation reaction of the initial carbon, a strongly exothermic reaction:

$$C + O_2 = CO_2 \tag{3}$$

then maintaining the temperature over time will be achieved by a continuous supply of carbon and oxygen, according to the scheme in Figure 3.

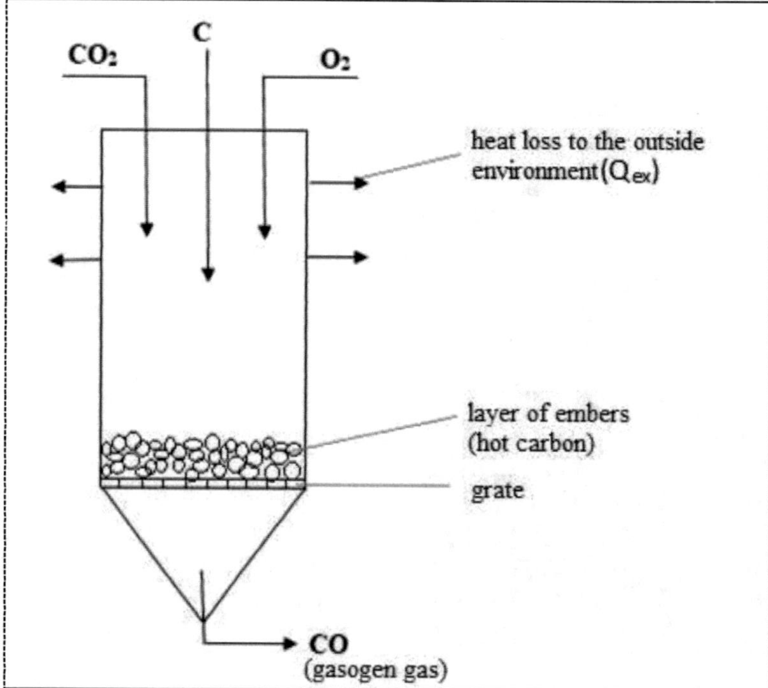

Figure 3. Scheme of operation of the gasification installation for the conversion of CO_2 to CO.

The carbon supply must be equal to the amount consumed by the reduction reaction, plus the amount that burns to produce the heat lost to the outside, corresponding to an adiabatic reactor. The amount of oxygen results from the carbon combustion reaction necessary to maintain the thermal regime. Figure 4 shows the carbon balance required for the operation of the installation.

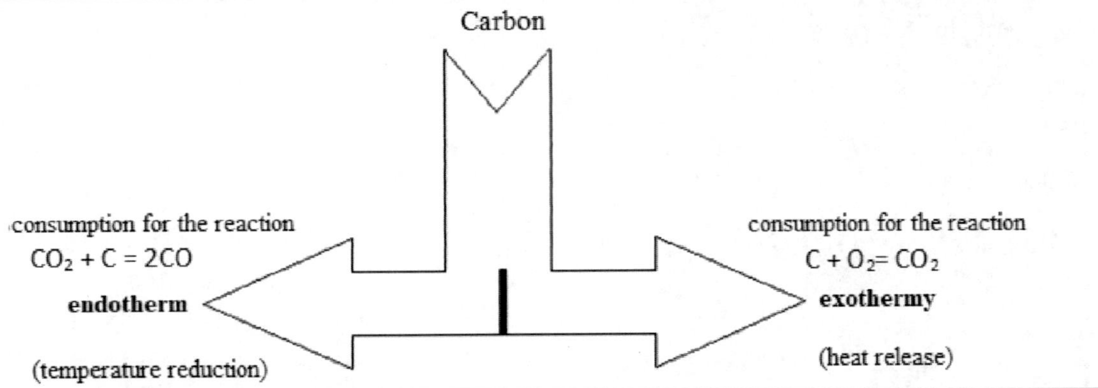

Figure 4. Carbon balance in the gasification installation.

The heat balance will lead to the equilibrium relationship:

$$Q_{exothermic} = Q_{ex} \text{ [kW]} \quad (4)$$

The endothermic reaction will lead to the reduction by the Δt_a value of the gasogenous gas (CO) temperature, according to the heat balance equation:

$$Q_{endothermic} = V_{CO} \cdot c_{CO} \cdot \Delta t \text{ [kW]} \quad (5)$$

Where V_{CO} is the volume of CO in m^3/s at the installation outlet;

c_{CO} – specific heat in kJ/m^3_NK;

Δt – temperature drop corresponding to the carbon dioxide reduction reaction.

It results a release of CO gas (gasifier gas) at 150-300°C, facilitating its energetic use.

The conversion degree of CO_2 to CO will depend on the carbon layer temperature. Table 1 shows the conversion of CO_2 to CO for temperatures between 450 and 950°C.

Table 1. CO_2 conversion rate as a function of temperature.

Temperature [°C]	CO [%]	CO_2 [%]
450	0.6	99.4
550	10.7	89.3
650	39.8	60.2
800	93	7
950	96	4

For conversion values above 90%, a temperature of the carbon layer in the range of 900 – 1000°C is necessary.

Figure 5 shows the 5kW gasification installation that can be used to transform CO_2 into CO.

IOP Conf. Series: Materials Science and Engineering 1290 (2023) 012014 doi:10.1088/1757-899X/1290/1/012014

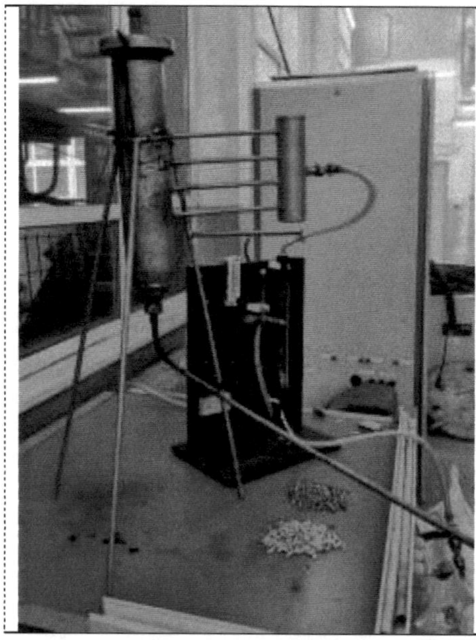

Figure 5. 5kW gasification installation equipped for CO_2 to CO conversion testing.

The energetic use of gasogenous gas can be achieved in an internal combustion engine (M), or in a heat generator (C), including by mixing with methane - Figure 6. The reduced calorific value for CO can be improved by mixing it with CH_4.

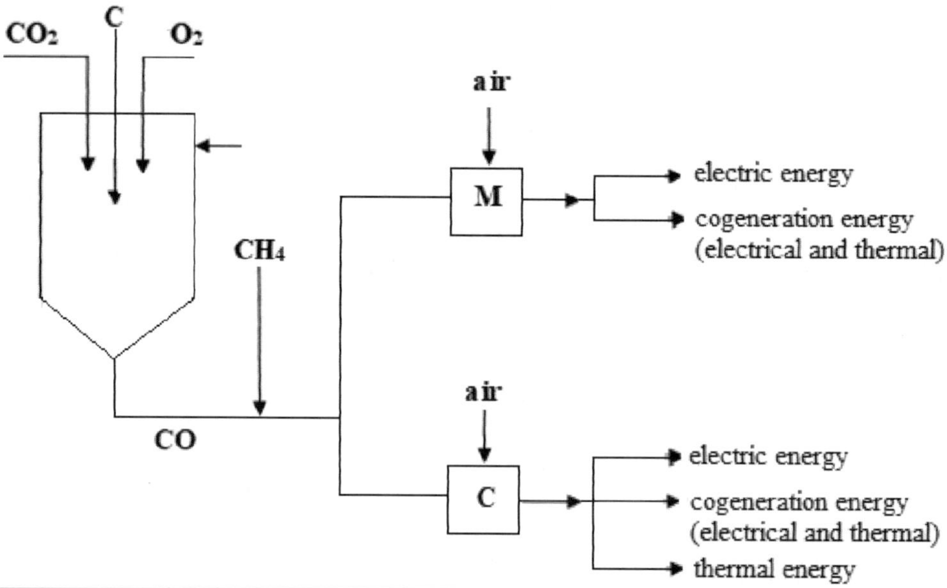

Figure 6. Scheme of energy production from CO.

For a unitary CO_2 to CO conversion degree, a neutrality occurs for the CO_2 circuit in a complex gasogenous installation – power production facility, according to the scheme in Figure 7.

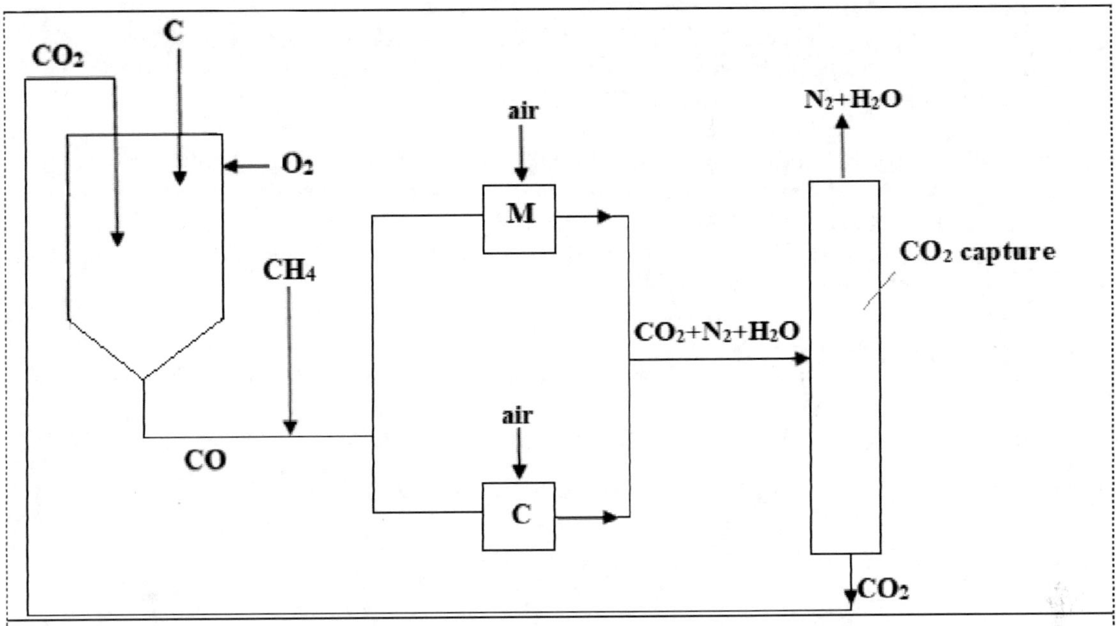

Figure 7. CO_2 recirculation scheme for its neutrality in energy production.

The experimental installation contains a 5kW adiabatic gasifier with fixed carbon bed. The composition of the gasifier gas will be determined, and the material and heat balance will provide data on the process efficiency.

4. Conclusions

The work includes a study on the new concept of using carbon dioxide as an energy vector. This will bring changes and new implications in its capture process, but especially in its storage [11], [12], [13].

Carbon dioxide can become an energy vector by converting it into methanol and carbon monoxide, fuels with real possibilities of energetic valorization. The paper considered the extension of research to the area of CH_3OH residential heating and CO production from CO_2 in a 5MW pilot installation. For the study of CH_3OH combustion at a residential installation, the experiments can be performed on the 55kW pilot boiler. Both experimental facilities are owned by the laboratory of Politehnica University of Bucharest.

The applied research on the CH_3OH combustion is a result of the innovative technology for obtaining methanol by using CO_2, a much more efficient technology than that involving carbon monoxide hydrogenation.

Residential heating with low CO_2 emissions can also be coupled with electricity production by electric generators, methanol replacing liquid hydrocarbons in both cases. Future research should be focused on global calculations of overall efficiency and total costs in order to complete the results presented in the current work.

Acknowledgments

This work was supported by a grant of the Ministry of Research, Innovation and Digitization, CNCS - UEFISCDI, project number PN-III-P4-PCE-2021-0777, within PNCDI III, contract PCE 5/2022.

References

[1] Zhang H, Wang L, Fortes M, Herle J, Marechal F and Desideri V. 2020 *Applied Energy* **258** 114071
[2] Bukhtiyarova M, Lunkenbein T, Kahler K and Schlogl R 2017 *Catal Lett.* **147** 416-427.
[3] Olah G A 2005 *Angew. Chem. Int. Ed.* **44** 2636–39.

[4] Perez-Fortes M, Schoneberger J C, Boulamanti A and Tzimas E 2016 *Appl. Energy* **161** 718-732.
[5] Swanson R, Statrio J and Brown R 2010 *Techno-Economic Analysis of Biofuels Production Based on Gasification* (*Technical Report, NREL: Golden*, CO, USA).
[6] Huber G, Iborra S and Corma A 2006 *Chem. Rev.* **106** 4044–98.
[7] Specht M, Staiss F, Bandi A and Weimer T 1998 *Int. J. Hydrogen Energy* **23** 387-396.
[8] Lazaroiu G and Mihaescu L 2021 *Innovative Renewable Waste Conversion Technologies* (Springer) ISBN 978-3-030-814304 Switzerland AG.
[9] Lazaroiu G and Mihaescu L 2018 *Biomasa solidă sursă de energie curată* (Ed. Politehnica Press) ISBN 978-606-515-838-2 Bucuresti.
[10] Lazaroiu G and Mihaescu L 2020 *Spre o energetică circulară* (Ed. Politehnica Press) ISBN 978-606-515-838-2 Bucureşti.
[11] Kuliyev S A, Aksongur S, Mat M D, Ibrahimoglud B and Kozlue MD 2009 *J. Electrochem. Soc.* **25** 1093–98.
[12] Goeppert A, Czaun M, Jones J P, Prakash G S and Olah G A 2014 *Chem. Soc. Rew.* **43** 7957–8194.
[13] Nieminem H, Laari Aand Koiranen T 2019 *Processes* **7** 405.

An analysis of the economic production and use of green hydrogen

G Lăzăroiu[1], L Mihăescu[2], G Negreanu[2], R M Grigoriu[1,3] and I Simion[1]

[1] University "Politehnica" of Bucharest, Department of Energy Generation and Use, 313 Splaiul Independenței, 060042 Bucharest, Romania
[2] University "Politehnica" of Bucharest, Department of Technical Thermodynamics, Engines, Thermal and Refrigeration Equipment, 313 Splaiul Independenței, 060042 Bucharest, Romania
[3] Author to whom any correspondence should be addressed

Email: manuela.grigoriu.1968@gmail.com

Abstract. The paper studies technological and economic details of green hydrogen production. Among all hydrogen categories defined by the ratio between production technology and CO_2 environmental pollution, for energy applications, only green hydrogen has a future. The same projection can be seen in the transport sector. An analysis of the H_2 thermochemical formation technologies showed that ecological hydrogen, called blue hydrogen, can be obtained only by a total separation of hydrogen from CO_2, to be recirculated. The thermochemical decomposition of water involves the use of three types of electrolysis cells, of which the solid acid cell (SOEC) is the most economical. Membrane-free water electrolysis is considered as a technology of the future. Production costs are related to storage, compression and distribution. The environmental tax through CO_2 emission was also included in the production cost calculation. A numerical application related to the green hydrogen production is also presented, as well as the achievements in field in EU and Romania, along with future perspectives.

1. Hydrogen production – green hydrogen

Hydrogen is a gas with totally different properties compared to other gaseous fuels (especially hydrocarbons). Thus, it consists of a proton and an electron. At the moment, since it is not freely found in nature, it is considered an energy vector, not being able to be included in the energy sources category.

Currently, hydrogen is mainly used in the oil industry, steel industry, energy production and transport; however, the achievements are still being researched and applied locally. The main challenge in the hydrogen uses refers especially to the economic aspect (costs).

Hydrogen can be produced by thermochemical, electrolytic and biological processes.

a. Thermochemical processes

They release hydrogen from organic materials based on the temperature-chemical reaction binomial. The heat used in closed chemical cycles produces hydrogen from water. For current technologies, natural gas reforming (CH_4) is the most used solution. The process is perfected by steam-methane reforming and partial oxidation. The steam-methane reforming reaction leads to final products CO and H_2:

$$CH_4 + H_2O(+heat) = CO + 3H_2 \quad (1)$$

The reaction is continued by gas-water conversion (exothermic reaction):

$$CO + H_2O = CO_2 + H_2 \quad (2)$$

Content from this work may be used under the terms of the Creative Commons Attribution 3.0 licence. Any further distribution of this work must maintain attribution to the author(s) and the title of the work, journal citation and DOI.
Published under licence by IOP Publishing Ltd

The reforming reaction requires a high temperature (700-1000 °C) and a pressure of 3 to 25 bar (the use of a catalyst is also imperative). Through adsorption and pressure variation, the hydrogen is separated from CO_2, becoming pure. Results similar to the hydrogen production technology by hydrocarbons reforming are also obtained by the partial oxidation technology:

$$CH_4 + 0.5O_2 = CO + 2H_2(+\text{heat});\ CO + H_2O = CO_2 + H_2(+\text{heat}) \qquad (3)$$

It is to be noted that partial oxidation has a lower hydrogen production than the reforming technology (3 m_N^3 hydrogen versus 4 m_N^3 when processing a $m_N^3 CH_4$). Gasification is a thermo-chemical technology for the transformation of organic materials at temperatures above 700 °C (with combustion controlled by the oxygen and/or steam amount) into a gas containing mainly CO, H_2 and N_2 (using air as a combustion medium). The carbon monoxide will later be subjected to the gas-water conversion reaction, which is the generator of H_2. Finally, through absorption processes or with special membranes, the hydrogen will be separated from the other gaseous components.

Gasification can also be used industrially from coal, to household waste, and even biomass. Pyrolysis is a thermal process of heating organic materials in the absence of air. The gaseous components are represented by CO, CO_2, H_2, CH_4 and N_2. Next, in order to obtain pure hydrogen, the gasification stages are undergone. The use of biomass for hydrogen production represents a first step in obtaining "green hydrogen"; this technology should not generate CO_2.

Another application of biomass in obtaining green hydrogen includes firstly its biological treatment, with ethanol or oils production, which are subsequently subjected to the thermochemical reforming process, similar to the processes from hydrocarbons. The advantage lies in the easier handling of liquid products than raw biomass. Reforming of biomass-derived hydrocarbons have larger molecules than fossil hydrocarbons and are still requiring testing in order to mature the process.

b. Thermochemical decomposition of water

The thermochemical decomposition of water leads to a green hydrogen production if the high temperature required by the process is based on solar, wind or nuclear energy. Temperatures must reach values between 500 and 2000 °C. This technology is completely carbon-free, as water breaks down into only hydrogen and oxygen.

Achieving green hydrogen will have to go hand in hand with developments in storage and transport pipelines. New reinforced thermoplastic pipes for pressures up to 42 bar are expected to be a viable solution.

At the current technological stage, hydrogen obtained by reforming hydrocarbons that generate a lot of CO_2 is significantly cheaper than hydrogen obtained by processes with low of CO_2 generation. The method of hydrogen production from water by the chemical electrolysis completely eliminates the generation of CO_2 (primary energy costs from nuclear or geothermal sources contribute significantly to reducing the production cost). Figure 1 shows the green hydrogen production scheme.

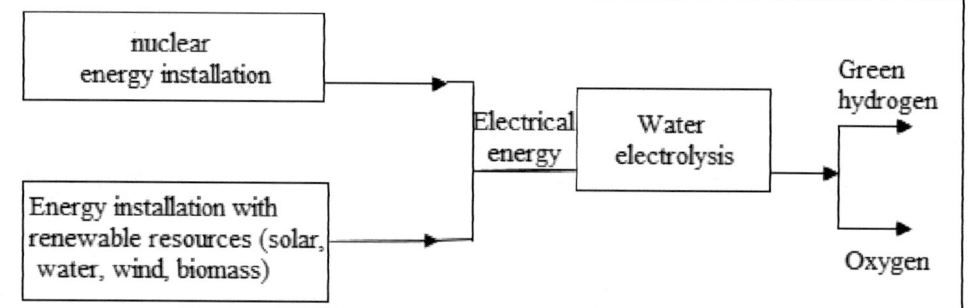

Figure 1. Green hydrogen production scheme.

Blue hydrogen is obtained through the process of reforming hydrocarbons and coal, a process accompanied by the capture of carbon dioxide.

Electrolysis technology (electrolysis cells) includes:

IOP Conf. Series: Materials Science and Engineering 1290 (2023) 012015 doi:10.1088/1757-899X/1290/1/012015

- Alkaline electrolysis cells (AEC);
- Proton exchange membrane electrolysis cells (PEMEC);
- Solid oxide electrolysis cells (SOEC).

SOEC technology is the most efficient technology among them.

Green hydrogen can use biogas as raw material, the process being steam reforming (SR), dry reforming (DR), dry oxidation reforming (DOR), partial oxidation reforming (POR) and autothermal reforming (ATR). Biomass gasification and pyrolysis constitute one of the most important sources for biogas generation.

Biological methods are also part of the category of renewable energy sources, but the necessary efficiency requires additional studies.

The aquatic hydrogen represents a new technology for obtaining green hydrogen. The process involves internal combustion in oil reserves or in deep oil sands, achieved by the injection of oxygen. In the soil areas with temperatures above 350 °C, heavy molecules break down, followed by interaction with water. Pure hydrogen, obtained as a result of the CO_2 component retention in the soil, is captured on the surface by using appropriate membranes in the extraction wells (current membranes are made of palladium alloys). Aquatic hydrogen technology eliminates bringing hydrocarbons to the surface, this type of green hydrogen representing the technology of the future.

The investment cost includes mainly the obtaining of oxygen and the wells construction, along with the compression and transport facilities for H_2.

Regarding the industrial electrolysers, it is to be mentioned that they are based on the use of alkaline membranes and proton exchangers. The two gaseous components, respectively H_2 and O_2, are also separated by using a membrane; the unused membranes are elements that degrade quite quickly.

Recently, membrane-less electrolysis technology appeared, the electrolysers being classified into „passive" and „active" category. This technology uses high current densities (1-3.5 A/cm^2) and achieves a hydrogen purity of 95-99.5%.

Schematically, the production of hydrogen in electrolysers without a membrane is shown in Figure 2. The energy consumption for hydrogen production by dissociating water depends on its alkalinity, in correlation with the current density. For a density of 0.5 to 3.5 A/cm^2, the electricity consumption varies in the range of 20-50 kWh/kgH_2.

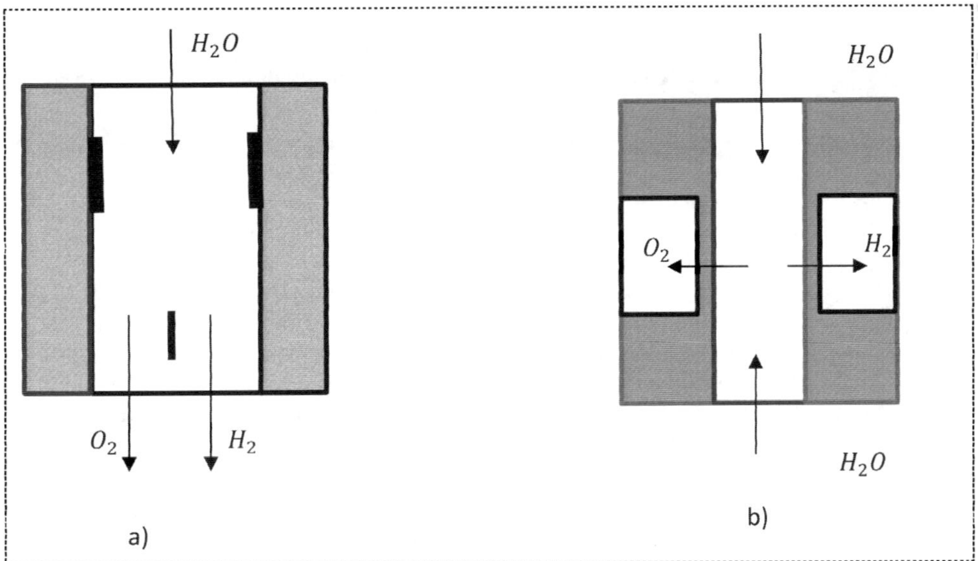

Figure 2. Scheme of electrolysers without membranes.
a) Active electrolyser; b) Passive electrolyser

Hydrogen purity is affected by temperature, current density, electrolyser geometry and flow conditions (expressed by the Reynolds number - Re). Figure 3 presents a graph of the influence of these functional parameters on hydrogen purity. It is to be noted the sensitive influence of the current density and less sensitive of the temperature level.

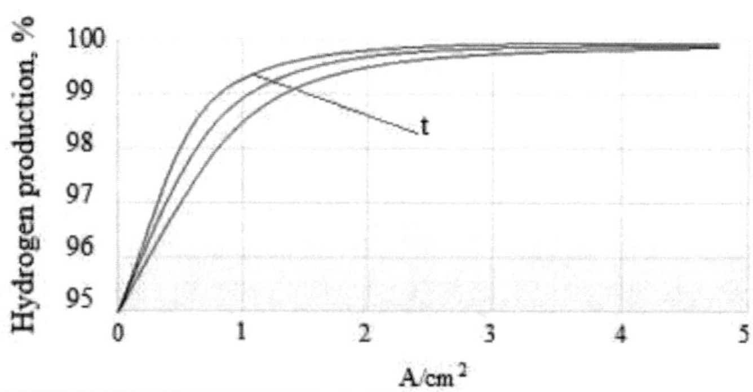

Figure 3. Hydrogen purity according to the functional parameters of electrolysers without membrane (Re=1500).

The cost of electricity represents the main operational cost. The annual operation and maintenance cost is usually more than 5% of the capital cost. Regarding the electricity cost, two options are taken into account: electricity from the national network and electricity from renewable energy sources. The second option is cheaper, but it is characterized by large power variations over time. A component of the costs includes the tax for the CO_2 emission; if the environmental tax is zero for green hydrogen production, in the case of hydrogen obtained by reforming hydrocarbons it becomes an important component in the total cost.

Hydrogen is also depending on the fuel cells operation. A big advantage of them lies in the capacity of cogeneration achieved by fractionation. Such installations produce both electricity and heat. Some issues are encountered in using hydrogen in the transport industry, but for residential consumption the advantages are obvious. Thus, thermal energy, sometimes perceived as a loss of efficiency, can be used for heating purposes [1].

The hydrogen production, including the green one, have to include other related operations, the main being the following [1]: storage, compression, distribution (transportation). Compression is imposed by both transport and storage operations. The final cost of hydrogen includes these related operations, but also those of purification and measurement (quantity, pressure, purity, etc.). The compression can be done by mechanical or non-mechanical process.

Metal hydride storage systems have proven effective for low to medium power plants. The storage of high flowrates includes grottoes from the salt pans and underground storage in porous medium.

In between, above-ground storage is mentioned, especially related to hydrogen production platforms.

Hydrogen injection into pipeline infrastructures for natural gas is currently studied intensively. The solution represents a shift towards the application of hydrogen in energy and industry. The proportion of hydrogen to methane is in an research phase (at the moment the proportion of up to 10% is preferred, but the experience gained through studies should help increase this percent). The reliability of the system will essentially depend on a static operation for flow and pressure (the pulsating variation of these influencing the lifetime of both new pipelines, but especially for those with a certain degree of ageing).

The injection of hydrogen into pipeline infrastructures for natural gas is full of study, the application solution of hydrogen in energy and industry. The proportion of hydrogen to methane is in a research phase.

At European Union level, most of these aspects are part of the NATURALHY project. The objectives of the project are to test the critical components of using hydrogen alongside natural gas [1].

2. The cost of hydrogen production

The development of hydrogen production technologies, intrinsically linked to those of use, are cost dependent. If at the beginning, in the phase of local hydrogen use for industrial purposes, the costs were dominated by those of production, the transition to using programs for energy, residential and transport purposes had widened the range of costs.

The current development stage proposes the use of black or grey hydrogen for industrial purposes, the future being represented by green hydrogen and its use in the energy-heating-transportation triad. The diagram in Figure 4 shows the amount of CO_2 resulting from hydrogen obtaining process depending on the production technology.

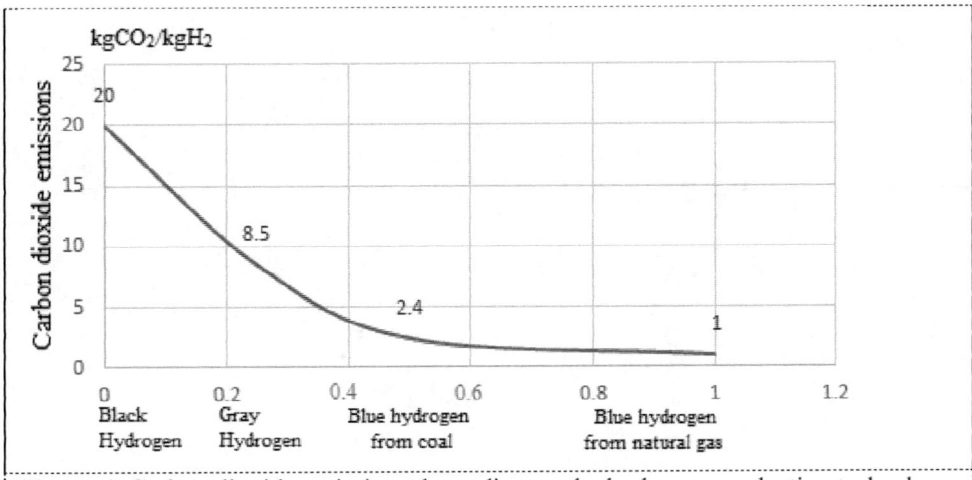

Figure 4. Carbon dioxide emissions depending on the hydrogen production technology.

In Figure 4 is also shown data for CO_2 emission from coal for blue hydrogen with CCUS (Carbon Capture, Utilization, and Storage) capture technology, for a CO_2 retention reactant of 90%.

Figure 5 shows the reported costs variation for obtaining hydrogen through various processes, compared to the cost of black hydrogen, without CCUS capture, considered as a unit.

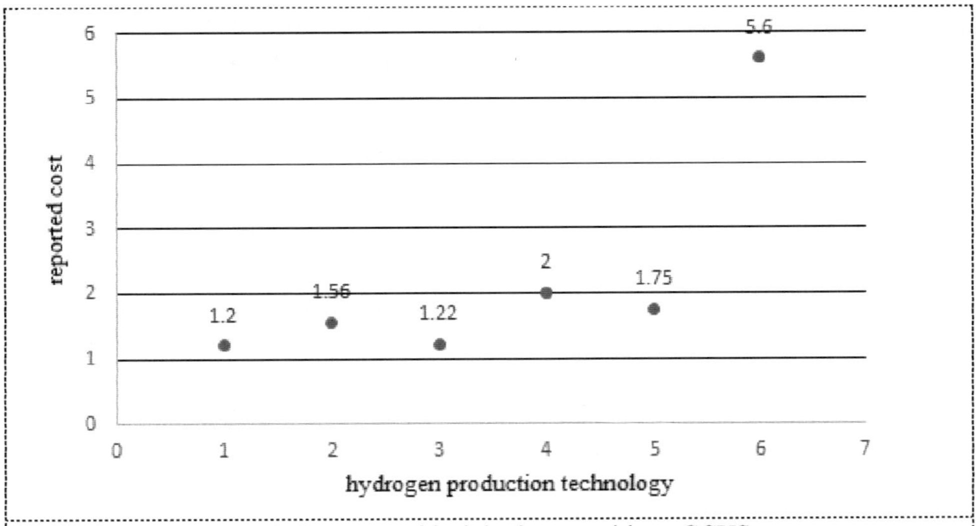

Figure 5. The costs variation versus black hydrogen without CCUS.

Regarding green hydrogen, its cost variation depends on the technology developed in various countries (Canada 1.15 – 5.64; Japan 1.95 – 5.22; Europe 1.75 – 5.18; Australia 3.65 – 5.67) [2], [3], [4], [5], [6].

Aquatic hydrogen technology, which is still under an immature study, leads to the lowest costs, about 0.23 compared to black hydrogen without CCUS [7], [8], [9]. The green hydrogen prices evolution at the current level of use is estimated over time in Figure 6.

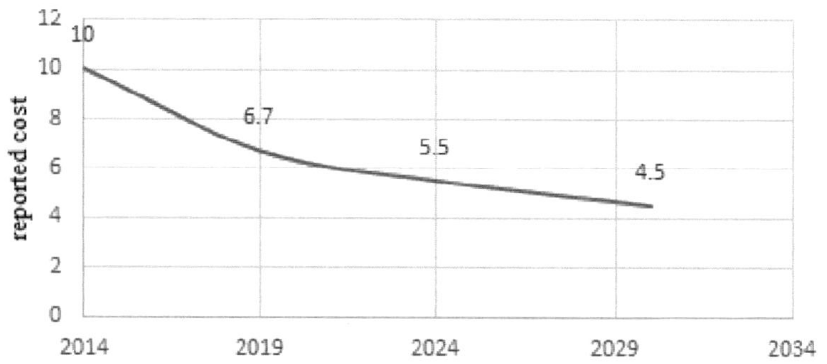

Figure 6. Green hydrogen price trends.

Most probably, due to the energy crisis triggered in 2022, some corrections will appear.

In the following paragraphs, an analysis of green hydrogen production costs will be carried out for a power of 1 MW at the level of the year 2021 (for the period after 2021, the price dynamics will contort the absolute values, but the trend will remain).

If a consumption of 4.3 $kWh/m_N^3H_2$ is considered, the following green hydrogen production will result (B_{H_2} in m_N^3/h)

$$B_{H_2} = \frac{1\times10^3}{4.3} = 230\ m_N^3/h \tag{4}$$

According to the data reported in [10], [11], [12], for a power of 1 MW, the investment cost **CI** is 2.2×10^6 Eu, and the operational value (cost) **CO** is 110,000 Eu/year (5% of the capital).
The production cost of the mass unit of green hydrogen will be (CP in Eu/kgH_2):

$$CP = \frac{CI+CO}{mH_2} \left[\frac{\text{Eu}}{\text{kgH}_2}\right] \tag{5}$$

where mH_2 – is the mass of hydrogen produced during the considered period of 15 years (a constant production was considered for a 290 days year, which corresponds to about 80% of the annual period). For the considered values, mH_2 will become:

$$mH_2 = B \times \rho_{H_2} \times 24 \times 290 \times 15 = 230 \times 0.1 \times 24 \times 290 \times 15 = 2.4 \times 10^6\ \ [\text{kg}] \tag{6}$$

For this mass of hydrogen, the green hydrogen cost in 2021 becomes:

$$CP = \frac{2.2\times10^6+15\times1.1\times10^5}{2.4\times10^6} = 1.51 \left[\frac{\text{Eu}}{\text{kgH}_2}\right] \tag{7}$$

The calculation was made for a conventional module for green energy production of 1 MW, but this calculation can be extended for higher powers. As already mentioned, from 2022, due to the energy crisis and inflation, the costs have undergone many changes that may ultimately lead to different results.

3. European achievements in the field of green hydrogen

The promotion of hydrogen in the European industrial area is carried out in particular by the Hydrogen Europe Association. It comprises 160 companies, 78 research organizations and 25 national business associations. Its main goal is to create a market for hydrogen-related technologies.

Currently, the most widespread uses of hydrogen are in the steel, chemical, glass and electronics industries. The petrochemical industry was associated to the chemical industry, including refining for cracking, desulfurization and synthesis for ammonia and methanol production. The share of hydrogen use is mainly defined by the ammonia industry (about 55%), followed by refineries (25%) and by methanol production (10%).

Regarding green hydrogen, in the European Union there are being currently implemented projects envisaging the development of electrolysers up to 1 GW power.

From a social point of view, various communities still do not share the same vision regarding the dynamics of innovative and technological aspects related to hydrogen. Overcoming this obstacle is one of the goals of the near future to achieve the decarbonization objectives until 2050. In the first phase, until 2024, the objective includes the installation of electrolysers with a power of at least 6 GW, ensuring a production of up to 10^6t of green hydrogen per year. Related to the second phase (2024-2030), it is worth mentioning that Romania participated in the HyUnder project, in which it carried out two case studies, namely [1] underground storage study and estimation of the energy potential that can be produced from renewable sources, as well as the infrastructure potential for hydrogen production through water electrolysis. The analysis of the first study showed that there are 38 closed salt mines and 33 active mines in Romania. The sustainable underground volume includes:

- 22 794 000 m^3 in old mines;
- 14 388 000 m^3 in active mines;
- 27 540 000 m^3 in caverns.

Potential use:

- 6 900 000 m^3 in old mines;
- 7 340 000 m^3 in active mines;
- 10 620 000 m^3 in caverns.

The amount of hydrogen that can be stored is enormous, if we take into account the storage pressure of 40-80 bar and the current achievements that comprise only storage volumes of tens of thousands of cubic meters.

The second study has not yet been completed and no data have been published.

Transport within the EU using fuel cell electric vehicles (FCEVs) is a short-term objective. For this reason, it is estimated that by 2030 between 9 and 13% of vehicles will be based on this technology. The logistical support will include about 5,000 hydrogen fuelling stations, containing about 2.6×10^6 tons of H_2 at the level of 2030 [10], [11], [12].

4. Conclusions

It is estimated that, for the next period, hydrogen production with CO_2 emission will still be used for industrial purposes. Instead, the transport, residential services and power generation sectors will rely only on green hydrogen use.

Using hydrogen at the national and European Union level first requires its injection into the methane transport pipelines.

However, it is noted that the citizens are not sufficiently informed and have not yet learned about the economic and social issues related to the growing share of hydrogen use. The policy of small but sure steps will certainly contribute to assimilating the possibilities of using hydrogen for different technical applications.

The distribution of hydrogen mixed with methane raises costs both for new pipelines (about 3×10^6 to 5×10^6 [Eu/km]) and for reconditioning the existing ones (about 0.25×10^6 to 1.2×10^6 [Eu/km]).

A criterion of green hydrogen production efficiency can be defined by relating the energy held by the hydrogen to the hydrogen production energy:

$$\eta_{H_2\,\text{green}} = \frac{\text{Hydrogen energy}}{\text{Hydrogen production energy}} \left[\frac{\text{kWh}}{\text{kWh}}\right]$$

The information presented in the paper confirm the ascending trend of both production and applications of green hydrogen, in order to finally achieve the decarbonization target in 2030.

Acknowledgments
This work was supported by a grant of the Ministry of Research, Innovation and Digitization, CNCS - UEFISCDI, project number PN-III-P4-PCE-2021-0777, within PNCDI III, contract PCE 5/2022.

References

[1] Ioan I and Dumitru C 2022 *Hydrogen and natural gas in Romania* Bucharest: Editura DIDACTICA SI PEDAGOGICA.

[2] Babatunde O and Amit K 2013 *Applied Energy* 428–440.

[3] Minli Y, Ke W and Harrie V 2021 *Int. J. of Hydrogen Energy* 21261–273.

[4] Maddy E, Benjamin I, Tahra J, Hoda T and Lucie S, „https://www.pembina.org/pub/hydrogen-primer," 9 July 2020. [Interactiv]. Available: https://www.pembina.org/pub/hydrogen-primer

[5] Babatunde O and Amit K 2011 *Energy* 6326–39.

[6] Babatunde O, Amit K and Marc S 2016 *Int. J. of Hydrogen Energy* 8755–76.

[7] Philipp-Matthias H, Severin R D, Thomas G, Detlef S and Martin R 2019 *Int. J. of Hydrogen Energy* 12733–747.

[8] Wouter N, Johannes R, André F and Herib B 2018 *Applied Energy* 617-639.

[9] Dia M, Ali K and Robbie M 2020 *Int. J. of Hydrogen Energ* 24125–145.

[10] Alessandro M, Emanuele Q, Antonino C, Stephen C, Massimo S and Francesco C 2022 *Energy Conversion and Management* 115156.

[11] Ankica K, Matej P and Doria M 2021 *Int. J. of Hydrogen Energy* 10016–35.

[12] David Jure Jovan and Gregor Dolanc 2020 *Energies* **13** 1-16.

IOP Conf. Series: Materials Science and Engineering 1290 (2023) 012016 doi:10.1088/1757-899X/1290/1/012016

Amine treatment application for carbon dioxide capturing from gas emissions of a hybrid power plant with a solid oxide fuel cell

Ivan Hristov Beloev[1], Antonina Andreevna Filimonova[2], Alena Yurievna Vlasova[2], Ruzina Farsilovna Kamalieva[3], Artem Sergeevich Vinogradov[2], Iliya Krastev Iliev[3]

[1]Department of Transport, "Angel Kanchev" University of Ruse, Ruse, Bulgaria.
[2]Department "Chemistry and Hydrogen Energy" Kazan State Power Engineering University, Kazan, Russia.
[3]Department of Heat, Hydraulics and Environmental Engineering, "Angel Kanchev" University of Ruse, Ruse, Bulgaria.

E-mail: aachichirova@mail.ru, vlasovaay@mail.ru, rruzzi@yandex.ru, teema@list.ru

Abstract. The transition to alternative electrical and thermal energy sources has opened up many promising areas for the hydrogen-containing fuel study. When assessing the environmental prospects of using a solid oxide fuel cell, it was found that carbon dioxide is obtained as a by-product, which must be captured and disposed of. Carbon dioxide is captured by physical absorption, chemisorption, adsorption, catalytic hydrogenation, membranes, enzymes under the electricity action. In this work, an experiment was carried out on the carbon dioxide capture by absorption, namely, amine-containing substances. In the experiment, both single-component and multi-component solutions were tested. The experiment was carried out on a laboratory setup. Based on the data obtained, a conclusion was made on the carbon dioxide capture by various classes of amines. Recommendations have been made for industrial scale amine purification. Compliance with these requirements will achieve good results in industrial carbon capture.

1. Introduction

A promising direction in the decarbonization policy is the use of high-temperature solid oxide fuel cells (SOFCs). Fuel cells are being actively developed in Japan, South Korea, and the USA. For example, the key element of Germany's national hydrogen strategy is the "National Innovation Programme Hydrogen and Fuel Cell Technology" [1].

Hybrid power plants, including fuel cells and microturbines, are being intensively developed by MHI (Mitsubishi Heavy Industries) with financial support from the Japanese government [2]. High-temperature fuel cells have an operating temperature of 650-1000 °C, which allows the use of inexpensive catalysts, various types of fuel (natural gas, hydrogen of various purification degrees, hydrogen-containing gases), and also have a relatively high efficiency [3]. Fuel cells can be combined with such power units as a steam turbine, gas turbine, microturbine to increase the efficiency of

Content from this work may be used under the terms of the Creative Commons Attribution 3.0 licence. Any further distribution of this work must maintain attribution to the author(s) and the title of the work, journal citation and DOI.
Published under licence by IOP Publishing Ltd

IOP Conf. Series: Materials Science and Engineering 1290 (2023) 012016 doi:10.1088/1757-899X/1290/1/012016

electricity generation [4,5]. It should be noted that gas emissions after a high-temperature fuel cell directly depend on the fuel used. When using natural gas, the composition of the exhaust gases will include carbon dioxide, water, hydrogen. To create an environmentally friendly power plant, carbon dioxide must be captured and utilized.

2. Literature review

Today, there are carbon capture, use, and storage (CCS/CCUS) technologies that reduce greenhouse gas emissions that can be applied throughout the energy system. CCUS technologies include:
- capturing carbon dioxide (CO_2) after fuel combustion or in industrial processes;
- transportation of CO_2 by ship or pipeline;
- use as a resource to create valuable products or services;
- carbon dioxide underground storage [6].

The industrial method of capturing carbon dioxide is based on the construction of special installations at industrial facilities. Capture is the most capital intensive part of CCUS, accounting for approximately 60-70% of all investments. In most existing commercial projects, carbon dioxide is captured using mechanical separators and liquid solvents, usually based on amines, which form a chemical bond with carbon dioxide. Regeneration is carried out by heating, the bond with the amine is broken, carbon dioxide is separated from the amines and the amines are recycled. Based on the technologies of capture, an analysis of literary sources was made.

The authors Koss Peter and Koch Michael proposed a method for flue gases purification from CO2 using a circulating ammoniated solution. An ammoniated solution is prepared from ammonia; a promoter is introduced to increase efficiency. The promoter is piperazine, carbonic anhydrase and its analogues. Regeneration of a solution enriched with carbon dioxide is carried out by changing the temperature and pressure [7]. The ammoniated solution has a good absorption capacity, and there is also a joint absorption with sulfur and nitrogen compounds. However, the regeneration of the ammoniated solution is accompanied by significant losses due to high volatility [8].

To date, the most advanced technology for capturing carbon dioxide is its absorption by monoethanolamine (MEA). The absorbing capacity of 1 mol of primary amines is 0.5 mol of carbon dioxide. A large number of studies have been carried out to determine the effectiveness of MEA for flue gas purification, but the stability of amine solutions and the possibility of complete regeneration are still relevant. Some sources describe compositions based on amine-containing compounds that are more stable with minimal regeneration costs [9]. When using MEA to capture CO_2, sulfur, nitrogen compounds and suspended solids must be removed from flue gases [10-13].

The use of secondary, tertiary and hindered amines shows the same efficiency as MEA. But they require less energy for regeneration and have a greater absorption capacity. In some studies, the authors suggest using a mixture of amines from different classes in different proportions. This will achieve a synergistic effect in capturing carbon dioxide. Also, capital and operating costs for the regeneration process are reduced, the selectivity of the absorbent solution is increased, the reaction rate with CO_2 is increased, and the interregeneration period is increased [14–15].

The compositions described in some patents include at least one alkali metal salt or hydroxide, a polyamine, an alkanolamine. The content of active substances in the composition does not exceed 30%. Water acts as a solvent. The regeneration of these solutions is accompanied by an increase in temperature. During operation, periodic replenishment with polyamine is required [16–17].

Joch Ralph and Schneider Rüdiger proposed a two-component mixture based on an amino acid and an amino acid salt. In terms of molar concentration, the free acid is present in the mixture in excess compared to the salt of the amino acid. This provides a variable basicity index of the solution, which can be achieved at different temperatures of the absorption and desorption process [18].

Wang Zhilong and Zhae Yanfeng proposed a composition based on an organic amine and a functionalized ionic liquid. The concentration of the absorbent solution is 20-40%. The purification process is based on physical and chemical absorption, which proceeds within a narrow temperature range [19].

IOP Conf. Series: Materials Science and Engineering 1290 (2023) 012016 doi:10.1088/1757-899X/1290/1/012016

Based on the analysis of literary sources, it can be concluded that amines are widely used for purification of flue gases from carbon dioxide. It is worth paying attention to the fact that a complex composition of the absorber is mainly used, and insufficient data on the efficiency of using monosubstances are presented. Therefore, the main goal of this study was to determine the efficiency of carbon dioxide capture by monosubstances with different concentrations. The experimental part of the work is devoted to comparing the absorption capacity of various absorbents on a laboratory setup.

3. Materials and methods

A feature of the process is the absorption of the gas flow by the entire volume of the absorbent-liquid with the formation of a solution. To conduct a laboratory experiment, an installation was assembled (Figure 1).

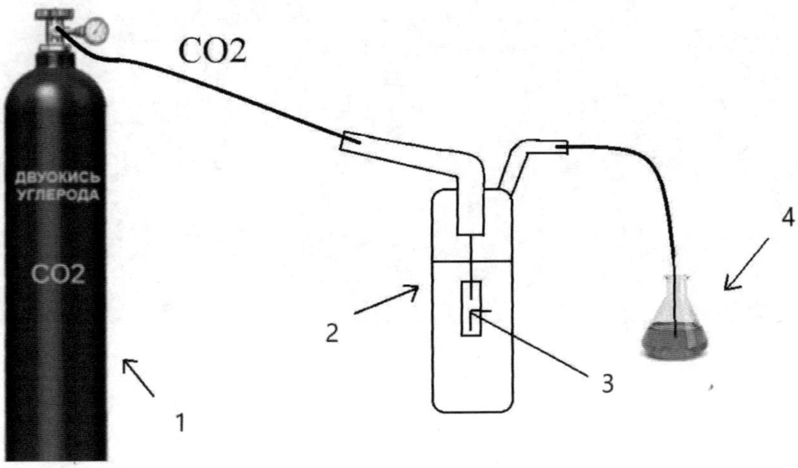

Fig.1. Scheme of laboratory installation of CO_2 absorption. 1 – CO_2 tank, 2 – an absorber, 3 – an air stone, 4 – a conical flask with an alkaline solution (solution receiver).

A CO_2 tank with a gas reducer was used as a source of carbon dioxide. With the help of a gas reducer, a gas flow rate of 0.032 l/min was set. The distribution of the gas supply to the absorber was provided by an air stone, which broke the gas bubbles into smaller ones. In this case, the contact of the gas phase and the absorbent is maximal. After the absorber, the gas passes into a conical flask with an alkaline solution of sodium hydroxide. The concentration of the sodium hydroxide solution was 0.05 N. The real image of the installation is shown in Figure 2.

Fig. 2. Laboratory setup.

This technique relies on double absorption, the first solution is an absorbent solution and the second is a sodium hydroxide solution. After absorption, carbon dioxide passes into a solution of sodium hydroxide, where bicarbonates are formed as reaction products. The quantitative determination of bicarbonates is carried out by titration with a solution of hydrochloric acid. This technique makes it possible to determine the total alkalinity and mass concentrations of carbonates and hydrocarbonates. A solution of HCl 1 mol-eq/ l is used as a titrant. Titration was carried out using phenolphthalein and methyl orange indicators.

4. Results and discussion

The purpose of the experimental part was to determine the absorption capacity of amine solutions of various concentrations. The simplest and cheapest absorbent solutions were also tested.

The following substances were used as absorbents: distilled water, monoemanolamine (MEA), diethanolamine (DEA), triethanolamine (TEA), dimethylethanolamine (DMEA), compounds with promoters and sodium hydroxide solution.

After purging the absorbent solution with carbon dioxide, the alkalinity of the receiver solution was determined by the titrimetric method (hydrate, carbonate and bicarbonate alkalinity). The results are shown in Table 1. These results quantify the carbon dioxide slippage after reaction with the absorbent. This factor is characterized by such indicators as carbonate and bicarbonate alkalinity.

Table 1. Results of the experiment

Absorbent	Solution concentration, %	Alkalinity, mg-eq/l

IOP Conf. Series: Materials Science and Engineering 1290 (2023) 012016 doi:10.1088/1757-899X/1290/1/012016

		Hydrate	Carbonate	Bicarbonate
Distilled water	-	0	0	60
MEA - 99%	5	36	24	0
	10	48	12	0
	15	54	6	0
MEA - 5% Piperazine -1%	-	8	52	0
DEA-100%	5	0	42	18
	10	0	60	0
	15	4	46	0
TEA-99%	5	0	20	40
	10	0	20	40
	15	0	44	16
DMEA-99%	1	0	28	32
	5	0	50	10
	10	0	56	4
Ammonia solution-25%	5	0	60	0
	10	12	48	0
	15	54	6	0
Sodium hydroxide -99%	5	44	16	0
	10	48	12	0
	15	60	0	0

Based on this table, the effectiveness of the studied absorbent solutions is clearly visible. The higher the hydrated alkalinity, the less carbon dioxide "leaked" into the receiver solution. Bicarbonate alkalinity indicates the amount of sodium hydroxide bound to CO_2, which indicates a large amount of carbon dioxide that has slipped through. Carbonate alkalinity indicates an intermediate stage of the reaction between CO_2 and NaOH.

Based on experimental data, calculations of the efficiency of the adsorbents used were carried out. The results are shown in Table 2. It is necessary to determine the amount of carbon dioxide supplied to the absorber for 8 minutes of purging. The volume of gas was 0.256 l (11.43 mmol). To calculate the amount of carbon dioxide that has slipped into the sodium hydroxide solution, we use the following formula:

$$\left(\frac{Al_c}{2} + Al_{bic}\right) * V_s, mmol \tag{1}$$

For calculations using this formula, it is necessary to use the data from Table 1, namely the concentrations of carbonate and bicarbonate alkalinity. The sample volume V_s was 0.1 l.

In order to determine the amount of carbon dioxide absorbed, it is necessary to subtract the amount of gas that reacted with the sodium hydroxide solution from the amount of carbon dioxide.

Table 2. Efficiency of the adsorbents.

Absorbent	Solution concentration, %	The amount of CO_2 absorbed by the sodium hydroxide solution (slip), mol	The amount of CO_2 captured by the absorbent, mmol	Efficiency, %
Distilled water	-	6	5.43	48
MEA-99%	5	1.2	10.23	90
	10	0.6	10.83	95
	15	0.3	11.13	97
MEA- 5% Piperazine -1%	-	2.6	8.83	77
DEA-100%	5	3.9	7.53	66
	10	3.0	8.43	74
	15	2.3	9.13	80
TEA-99%	5	5.0	6.43	56
	10	5.0	6.43	56
	15	3.8	7.63	67
DMEA-99%	1	4.8	6.63	58
	5	3.5	7.93	69
	10	3.2	8.23	72
Ammonia solution -25%	5	3.0	8.43	74
	10	2.4	9.03	79
	15	0.3	11.13	97
Sodium hydroxide -99%	5	0.8	10.63	93
	10	0.6	10.83	95
	15	0	11.43	100

Analyzing the results obtained, it can be seen that the following absorbent solutions have the best absorption capacity: MEA-15%, ammonia solution-15%, sodium hydroxide solution - 15%.

Features of the use of MEA involve the use of equipment for regeneration. When regenerating amines, it should be taken into account that they undergo degradation. Sulfur and nitrogen compounds have a particularly negative effect on MEA. After regeneration, the absorption capacity decreases and a fresh solution needs to be added. To date, there are no technologies that would ensure the complete regeneration of amines. Technological difficulties and the need for constant addition of a fresh absorbent solution limit the applicability of this method on an industrial scale.

The use of an ammonia solution is also accompanied by difficulties in the regeneration process, and is also complicated by a strong, pungent odor. Therefore, the use of amines to capture carbon dioxide is a rather controversial issue. In terms of absorption efficiency, sodium hydroxide solution is no less effective than amine-containing solutions.

The use of sodium hydroxide solution in the technological process has a number of advantages: simplicity of design and ease of maintenance, high economic efficiency of the process, cyclical process (repetitiveness), safety of personnel and the environment. Based on the results of the experiments, a technology for capturing carbon dioxide with an alkali solution with subsequent regeneration is proposed. The technological scheme includes an absorber, a contact container, a filter press and a reagent facility, which is presented in the form of tanks for solutions preparing (Figure 3).

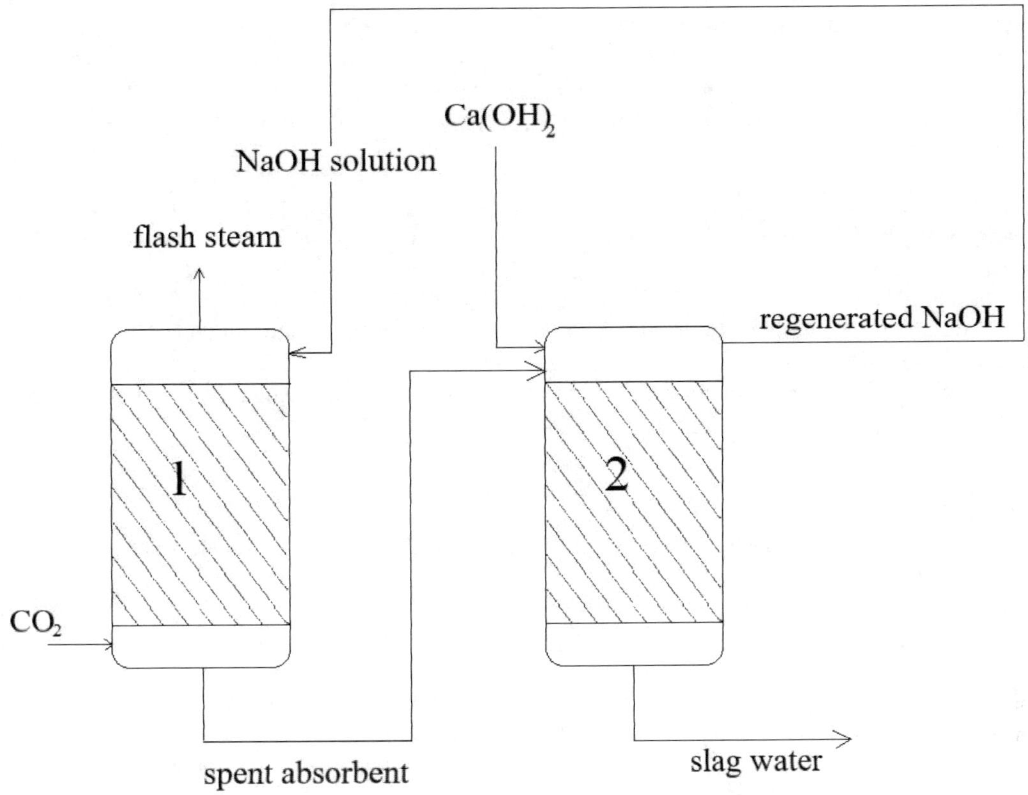

Fig.3. Technological installation for carbon dioxide capture and utilization. 1 – absorber, 2 – contact container,

The key feature of this technology is cyclicity and the absence of additional equipment to maintain temperature and pressure parameters. Carbon dioxide enters the absorber. In the top of the absorber there is a distribution device that increases the contact of the phases. The CO_2 flow moves in the opposite direction to the direction of the absorbent solution. The absorbent solution is saturated with sodium carbonate and bicarbonate. The saturated waste solution enters the regeneration, where it contacts with a lime solution. As a result of the reaction, a liquid phase is formed - a regenerated sodium hydroxide solution and calcium carbonate, which precipitates in the form of a solid phase. The regenerated absorbent solution is fed back into the absorber.

The solid gypsum product that is formed during the regeneration process can be used as a commercial product, or in any other industry. This method of capturing and utilizing carbon dioxide is simple and economically available.

5. Conclusion

The transition to hydrogen energy does not imply the absence of carbon dioxide emissions into the atmosphere. For solid oxide fuel cells, either natural gas or hydrogen-containing gases can be used as fuel. Therefore, carbon dioxide is released during the process, which must be captured and disposed of, for example, using the absorption method, where amine solutions act as absorbents. The paper presents the results of the absorption capacity of various amines, as well as their compounds.

The experimental part of the work was to create a laboratory absorption unit. A number of absorbent solutions were tested on this laboratory setup. Based on the results obtained, the absorption capacity was calculated and the most effective absorbents were identified: MEA-15%, ammonia solution-15%, sodium hydroxide solution - 15%.

The use of amine-containing absorbents is complicated by the high degree of degradation of amines during regeneration. Therefore, a technological scheme was proposed for capturing and utilizing carbon dioxide based on the use of sodium hydroxide solution. Sodium hydroxide solution has a better absorption capacity than amines, and regeneration is carried out with an available chemical reagent (lime).

Acknowledgment

The results were obtained with the financial support of Ministry of Science and Higher Education of the Russian Federation "Study of processes in a fuel cell-gas turbine hybrid power plant" (project code: FZSW-2022-0001).

This work has been accomplished with the financial support from the grant number BG05M2OP001-1.002-0011-C02, financed by the Bulgarian Operational Programme "Science and Education for Smart Growth" (SESG) (2014-2020), and co-financed by the European Union through the European Structural and Investment Funds (ESIF).

.

References

[1] The National Hydrogen Strategy. Federal Ministry for Economic Affairs and Energy, Germany, Berlin, June 2020, 28P.

[2] Kobayshi Y., Tomida K., Nishiura M., Hiwatashi K., Kishizawa H., and Takenobu K. Development of Next-Generation Large-Scale SOFC toward Realization of a Hydrogen Society. Mitsubishi Heavy Industries Technical Review, vol. 52(2), pp. 111-116, 2015

[3] Ivanov-Shits A.K. and Murin I.V. Ionics of the solid body. St. Petersburg: Publishing House of St. Petersburg University. In 2 volumes. Volume 2. 2010.

[4] Yarmolchik Yu.P., Shrager R., Haberfelner, Pichler M., Kostich D., and Moroz G.V. Combined combustion of various industrial waste streams in boiler furnaces. Part 1, Energy. Proceedings of Higher Educational Institutions and Energy Associations of the CIS, vol. 63(3), pp. 236-252, 2020 (in Rus.)

[5] Wejrzanowski T., Ćwieka K., Skibinski J., and Lysik A. Microstructure Driven Design of Porous Electrodes for Molten Carbonate Fuel Cell Application: Recent Progress. International Journal of Hydrogen Energy, vol.45(47), pp. 25719–25732, 2022

[6] Nomoto H., Itoh M., Brown W., Fetvedt J., and Sato I. Cycle and Turbine Development for the Supercritical Carbon Dioxide Allam Cycle. Proceedings of the International Conference on Power Engineering, 2015

[7] Patent RU(11) 2 540 613(13) C2 Process gas cleaning system

[8] Ma S., Song H., Wang M., Yang J. and Zang B. Chem. Eng. Res. Des. vol. 91(7), p. 1327, 2013

[9] Zaman M. and Jay Hyung Lee. Invited review paper Carbon capture from stationary power generation sources: A review of the current status of the technologies. Korean J. Chem. Eng., vol. 30(8), pp. 1497-1526, 2013. DOI: 10.1007/s11814-013-0127-3

[10] Bello A.and Idem R.O. Ind. Eng. Chem. Res., vol. 45(8), p. 2569, 2005

[11] Goff G.S. and Rochelle G.T. Ind. Eng. Chem. Res., vol. 45(8), p. 2513, 2005

[12] Lawal A.O. and Idem R.O. Ind. Eng. Chem. Res., vol. 45(8), p. 2601, 2006

[13] Supap T., Idem R., Tontiwachwuthikul P. and Saiwan C. Ind. Eng. Chem. Res., vol. 45(8), p. 2437, 2005

IOP Conf. Series: Materials Science and Engineering 1290 (2023) 012016 doi:10.1088/1757-899X/1290/1/012016

[14] Rinker E.B., Ashour S.S. and Sandall O.C. Ind. Eng. Chem. Res., vol. 39(11), p. 4346, 2000
[15] Xiao J., Li C.-W. and Li M.-H. Chem. Eng. Sci., vol. 55(1), p. 161, 2000
[16] Patent RU(11) 2 589 166(13) C1 Russian Federation (51) IPC B01D 53/00 (2006.01)
[17] Patent RU (11) 2015 101 990(13) A Russian Federation (51) IPC B01D 53/14 (2006.01)
[18] Patent RU(11) 2 488 429(13) C2
[19] Patent RU (11) 2 600 348(13) C2 Russian Federation (51) IPC B01D 53/14 (2006.01)

IOP Conf. Series: Materials Science and Engineering 1290 (2023) 012017 doi:10.1088/1757-899X/1290/1/012017

The impact of cogeneration plant and a storage tank in a district heating system

D P Hiris[1], M C Dudescu[1], A Pocola[1] and M C Balan[1]

[1]Technical University of Cluj-Napoca, Bd. Muncii 103-105, 400641, Cluj-Napoca, Romania

E-mail: mugur.balan@termo.utcluj.ro

Abstract. The goal of the study is to investigate the thermal behavior and the benefits of a cogeneration system based on two gas fired engines and of a thermal storage tank in a biomass based district heating system situated in Romania. The investigation is based on an hourly analytical model of all the district heating components: biomass boiler, natural gas cogeneration engines, storage tank and consumers. The climatic data of the location are used as input in the model. The gas cogeneration engines coupled with the heat storage tank increase the flexibility of the system and allow the production of electricity even in periods with little or no heat demand. The analytical modelling of the storage tank considering the water stratification allows the correct sizing of the storage tank, in the planning stage of the project. The total yearly heat production is divided between 81.54 % produced by the proposed cogeneration plant and 18.46 % produced by the existing biomass boiler.

1. Introduction

The coupling between the combined heat and power (CHP) plant with thermal engines with a large water storage tank (WST) allows to increase the flexibility of the system. The thermal energy storage removes the operation of the engine on partial load with many starts and stops and has a positive effect on the performance and lifetime of the system [1]. If the CHP is coupled with a WST the economic efficiency can be improved allowing to run the engines when the price of the produced electricity is high, even if there is not enough thermal energy consumption in the same period [2].

The behavior of the CHP coupled with WST was investigated with specialized software like PSCAD [3], TRNSYS [4], Ebsilon [5,6], and methods developed through mixed-integer linear programming (MILP) [7,8], or MATLAB [6,9–12].

Investigated systems are placed in different countries from Europe and Asia like Italy [13,14], Germany [11,12,15], Belgium [16], Spain [4], Poland [8], Finland [17], and China [6,18].

Different types of CHP technologies was analyzed: Solid Oxide Fuel Cell [3], Proton Exchange Membrane Fuel Cell [10], coal-fired power plants [5,8,18,19], gas engine [4,15,20], natural gas combined with wood gas [12], gas-turbine coupled with a heat recovery steam generator [13], and biomass fired [7].

Most studies refer to district heating but some investigations approach small-size systems such as residential houses [9,20] or domestic hot water [10]. In the context of different sizes of heating systems, the volume of WST can vary from hundreds of liters [9,10] to tens of cubic meters [4], to hundreds of cubic meters [14], or to thousands of cubic meters [7,11,13].

Content from this work may be used under the terms of the Creative Commons Attribution 3.0 licence. Any further distribution of this work must maintain attribution to the author(s) and the title of the work, journal citation and DOI.
Published under licence by IOP Publishing Ltd

IOP Conf. Series: Materials Science and Engineering 1290 (2023) 012017 doi:10.1088/1757-899X/1290/1/012017

Tables 1 and 2 synthesize the main parameters investigated in similar studies available in the literature. The characteristic parameters of the CHP plant and of the WST are presented in table 1 and the characteristics of the simulation, of the consumers, and of the thermal network are presented in table 2.

Table 1. The characteristic parameters of CHP plant and WST in the technical literature.

References	CHP					WST								
	Q_{tp}	P_{tp}	Q_{elp}	P_{elp}	OT	Q_{ss}	P_{ss}	Q_{st}	P_{st}	TT	MT	SV	SC	Q_l
[3]		✓					✓				✓			
[9]		✓								✓				
[8]		✓		✓		✓		✓				✓	✓	
[10]				✓						✓	✓			
[4]			✓			✓		✓						✓
[20]	✓				✓									
[15]		✓					✓		✓				✓	
[16]		✓		✓									✓	
[17]		✓		✓			✓		✓					
[11]	✓					✓		✓		✓				
[12]	✓	✓				✓		✓						
[18]	✓	✓					✓		✓				✓	
[13]			✓			✓	✓							
[7]							✓		✓					
[6]		✓		✓				✓					✓	
[14]		✓					✓							
[19]										✓				
This study	✓	✓	✓	✓		✓	✓	✓	✓	✓	✓	✓		✓

Q_{tp} - Thermal energy production of CHP; P_{tp} - Thermal power production of CHP; Q_{elp}- electrical energy production of CHP; P_{elp} -electric power production of the CHP; OT - operating time; Q_{ss} - Thermal energy extracted from the tank; P_{ss} -Thermal power extracted from the tank; Q_{st} - Thermal energy introduced into the tank; P_{st} - Thermal power introduced into the tank; TT - Temperature of the water; SV - Storage volume; SC – Energy storage capacity; Q_l – Heat losses.

IOP Conf. Series: Materials Science and Engineering 1290 (2023) 012017 doi:10.1088/1757-899X/1290/1/012017

Table 2. The characteristics of simulation, consumers, and thermal network in the technical literature.

References	**τ [h]**	**Δτ [min]**	**Consumers**				**Thermal network**		
			Q_d	P_d	Q_{el}	P_{el}	ST	RT	HL
[3]				✓		✓	✓		
[9]				✓					
[8]	8760			✓					
[10]	1.166								
[4]	8760	6	✓						
[20]	8760	60	✓	✓					
[15]	8760	60							
[16]	8760	60			✓				
[17]				✓					
[11]	8760	3							
[12]	8760	15	✓	✓			✓	✓	
[5]						✓	✓	✓	
[13]	192720	60		✓			✓		✓
[7]					✓				
[6]	3624	60					✓	✓	
[14]	8760	15		✓		✓	✓	✓	
[19]				✓					
This study	8760	10	✓	✓	✓	✓	✓	✓	✓

τ – Duration of the simulation; Δτ – Time step; Q_d - Heat demand; P_d - Thermal power demand; Q_{el} - Electrical energy demand; P_{el} - Electrical power demand; ST - Supply temperature; RT - Return temperature; HL - Heat losses.

The goal of the study is to investigate the thermal behavior and the effects of a WST integrated in an existing district heating system with biomass boilers (BB), where a CHP plant based on gas engines is to be also integrated. The base thermal energy demand is to be covered by the CHP plant and the high thermal energy demand is to be covered by the BB.

2. Material and method

The investigated system is situated in Vatra Dornei, in Romania. The existing heat source is represented by two wood BB of 6 MW each, and it is proposed to add a CHP plant with 2 x 1500 kW electric power and 2 x 1600 kW thermal power, natural gas fired engines. It was also proposed a WST of 300 m^3. This size was determined following several consecutive iterations, considering that the storage volume must be convenient to allow both regimes of heating and of cooling, mainly in the summer period when the heat load is small, and a single engine is used.

The scheme of the system is presented in figure 1.

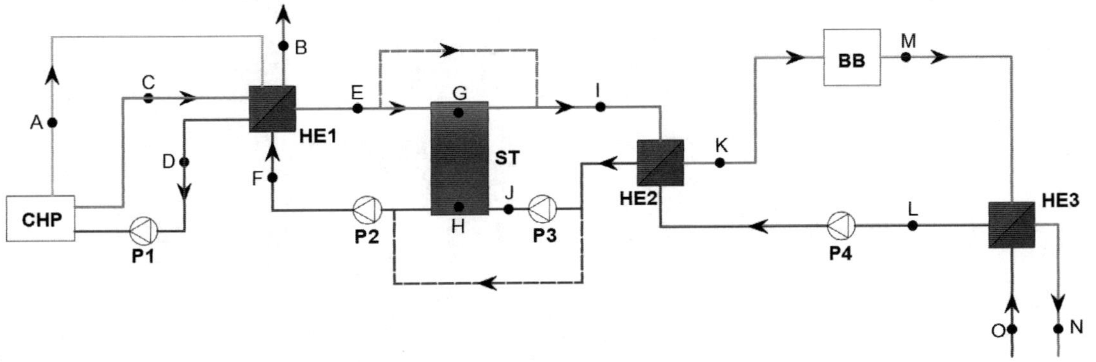

Figure 1. The scheme of the system.
CHP – Cogeneration system (2 x 1500 kW_{el}); ST – WST (300 m^3);
BB – Biomass boilers; HE1, HE2, HE3 – Heat exchangers; P1, P2, P3, P4 – Recirculation pumps.

The chosen nominal temperatures in the characteristic states of the system are presented in table 3.

Table 3. Nominal temperatures in the characteristic states [°C].

Position	A	B	C	D	E	F	G	H	I	J	K	L	M	N	O
Winter	374	120	82	70	87	J	min 75	J	G	L+5	I-5	O+5	N+10	f(t_{ex})	50
Summer	374	120	82	70	87	J	min 75	J	G	L+5	I-5	45	N+10	f(t_{ex})	10

The nominal temperatures are different in summer and in winter, because in summer when only domestic hot water (DHW) is required, the provided temperature of the thermal agent can be lower than in winter when heating is also required, and the heating systems of the consumers is of high temperature, based on high temperature radiators. The temperature of the thermal agent was considered dependent by the ambient temperature (t_{ex}). At its turn, the ambient temperature was considered from the available typical meteorological year (TMY) for Vatra Dornei and is presented in figure 2.

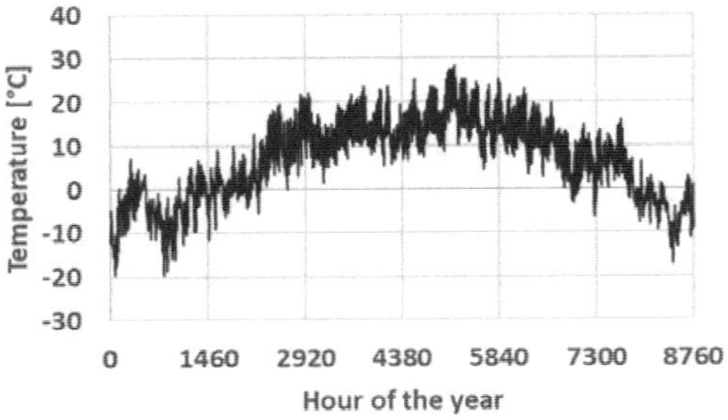

Figure 2. The ambient temperature from TMY.

IOP Conf. Series: Materials Science and Engineering 1290 (2023) 012017 doi:10.1088/1757-899X/1290/1/012017

The dependance of the flow temperature for the thermal agent, by the ambient temperature, is presented in figure 3 and the flow temperature during the year is presented in figure 4.

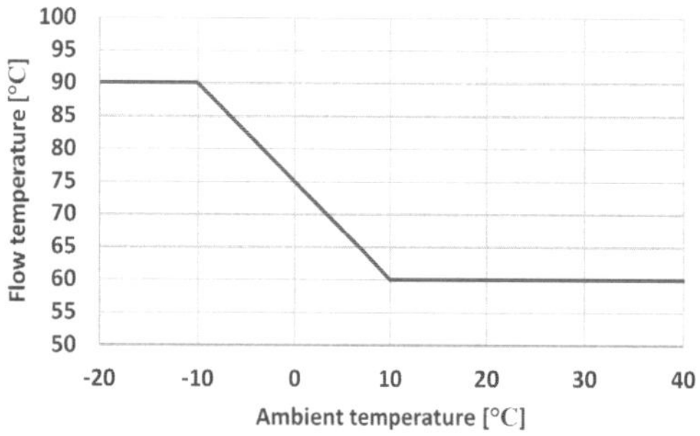

Figure 3. Flow temperature (state N) variation vs ambient temperature.

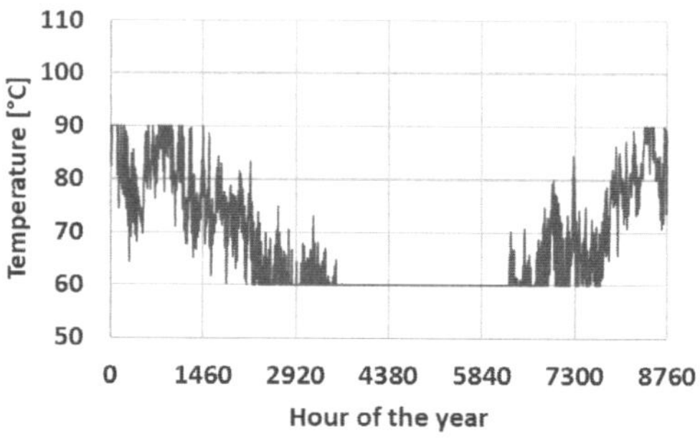

Figure 4. Flow temperature variation during the year.

The heat load was determined based on the available historical data. Figure 5 presents the historical and the calculated monthly heat load for both heating and DHW, and figure 6 presents the hourly variation of the heat load for both heating and DHW.

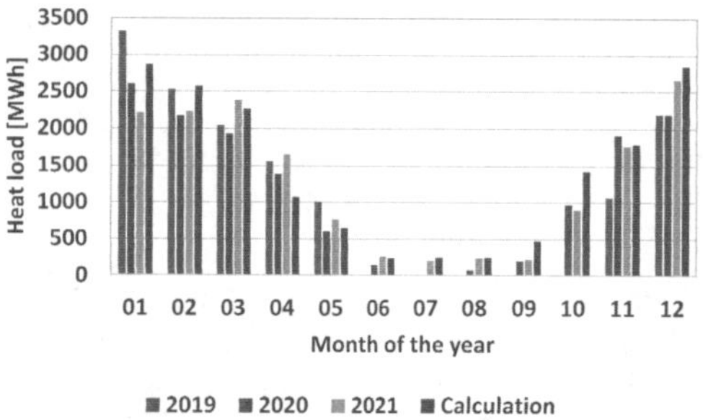

Figure 5. Monthly heat load.

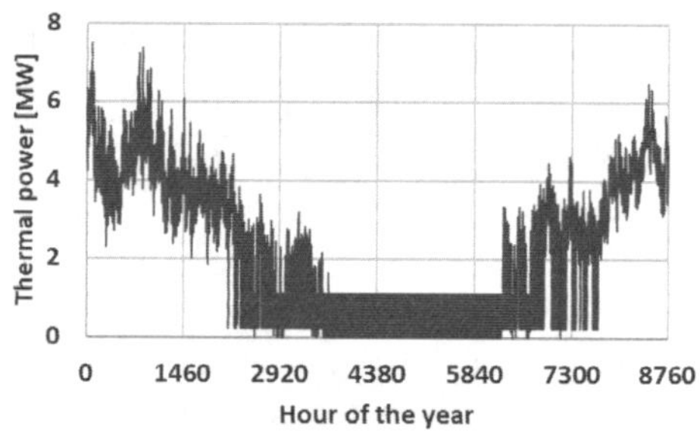

Figure 6. Hourly heat load.

The calculations to evaluate the thermal behavior of the WST and of all the system, were performed by simulation, using the analytical method. The WST was considered stratified and the temperature evolution inside the WST was determined based on thermal balance equations for each control volume. The computations method related to the WST is similar with the one presented in some previous studies [21–23]. The model of the WST is based on the 1D heat equation, that was solved by the finite difference method.

The discretization scheme and the basics of the mathematical model are presented in figure 7.

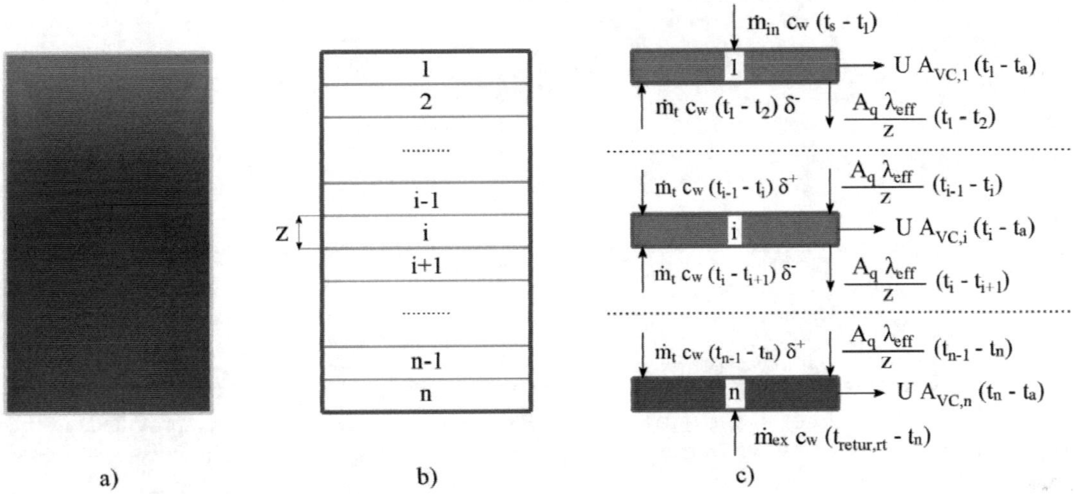

Figure 7. The discretization scheme and the basics of the mathematical model
a) The real tank; b) The discretized tank; c) The mathematical model.

The WST with the diameter of 8.6 m and the heigh of 5.16 m, was divided into 10 control volumes with the similar heigh of 0.516 m.

3. Results

The simulation computed the hourly variation of several parameters such as: temperatures and flow rates in all the characteristic states of the district heating system, temperatures distribution inside the WST, thermal and electric power provided by the CHP engines, thermal power provided by the BB, thermal power transferred through the heat exchangers, thermal power introduced in and extracted from the WST. Some of these results are presented as follows.

The temperature variation at the outlet of the WST through HE2 (state I) is presented in figure 8 and the temperature variation at the outlet of the HE2 through the BB (state K) is presented in figure 9.

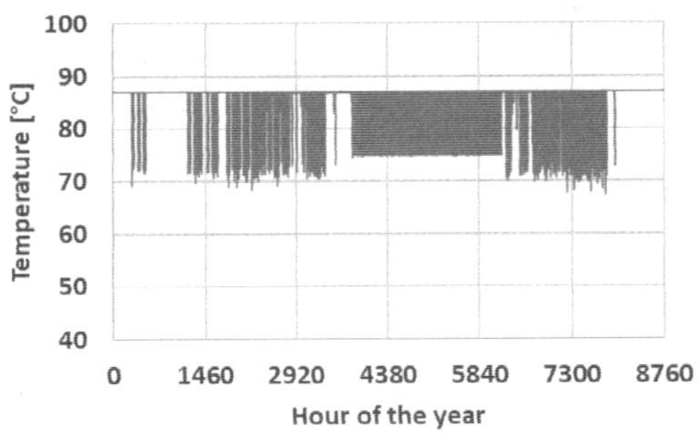

Figure 8. The temperature variation at the outlet of the WST through HE2 (state I).

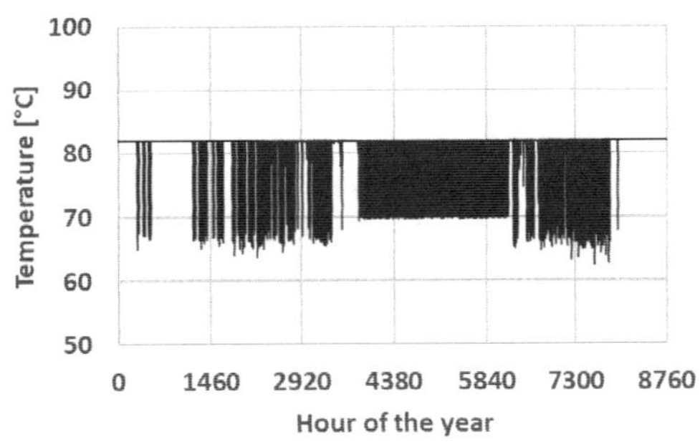

Figure 9. The temperature variation at the outlet of the HE2 through the BB (state K).

The typical comparison between the thermal power provided by the CHP engines and the thermal power consumed by the consumers of the district heating, both in winter and in summer conditions, are presented in figures 10 and 11 respectively.

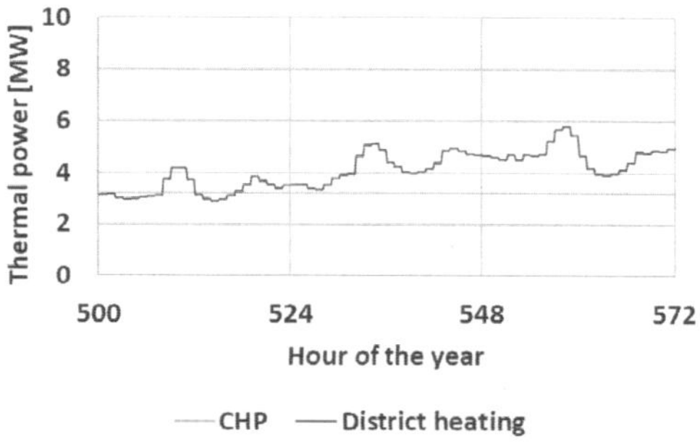

Figure 10. The thermal power provided by the CHP plant and consumed by the district heating in winter (3 consecutive typical days).

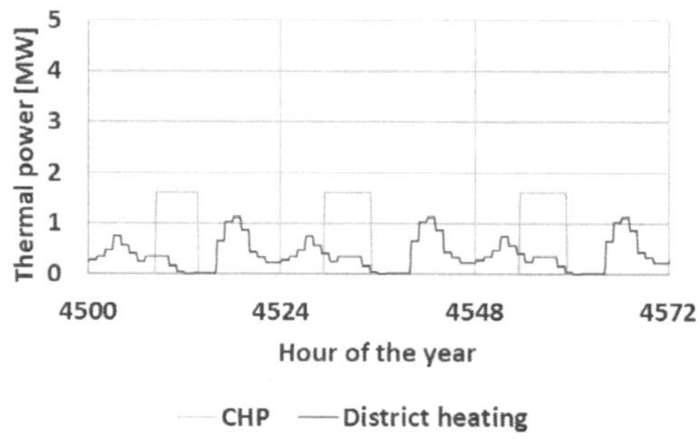

Figure 11. The thermal power provided by the CHP plant and consumed by the district heating in summer (3 consecutive typical days).

In winter, most of the time the required heating power is higher than the thermal power that can be provided by the CHP engines. In these conditions the BB will operate to provide the maximum required heating power and the WST will not be used because it will be rapidly discharged and most of the time it will be impossible to heat it. In winter, the WST will be bypassed. In summer, the thermal power of even a single CHP engine is higher than the required DHW heating power. In these conditions only a single CHP engine will be operated, and the WST will be used to increase the flexibility of the system with the major benefit that the engine can work mostly in periods with the high price of the provided electricity.

A typical temperature variation inside the WST control volumes, in 3 consecutive typical days of summer, is presented in figure 12.

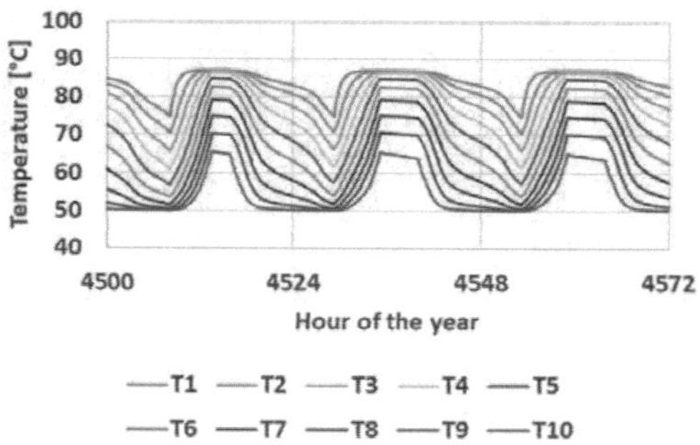

Figure 12. Typical temperature variation inside the WST control volumes, in 3 consecutive days of summer.
T1...T10 – The temperatures in the control volumes (numbered from top to bottom).

The upper layer temperature is of maximum 87 °C, resulted from the CHP engine. The lower layer temperature is of minimum 50 °C (the minimum return temperature of the thermal network), and of maximum 65 °C (the maximum accepted temperature to cool down the CHP engines).

IOP Conf. Series: Materials Science and Engineering 1290 (2023) 012017 doi:10.1088/1757-899X/1290/1/012017

The temperature distribution inside the WST in the charging (heating) period is presented in figure 13 and in the discharging (cooling) period is presented in figure 14.

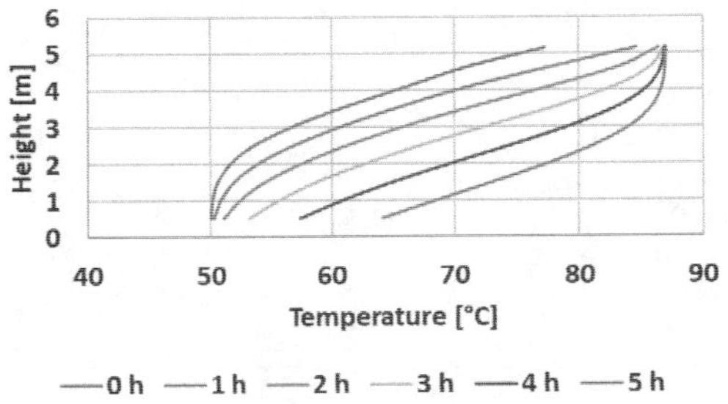

Figure 13. The temperature distribution inside the WST in the charging (heating) period.

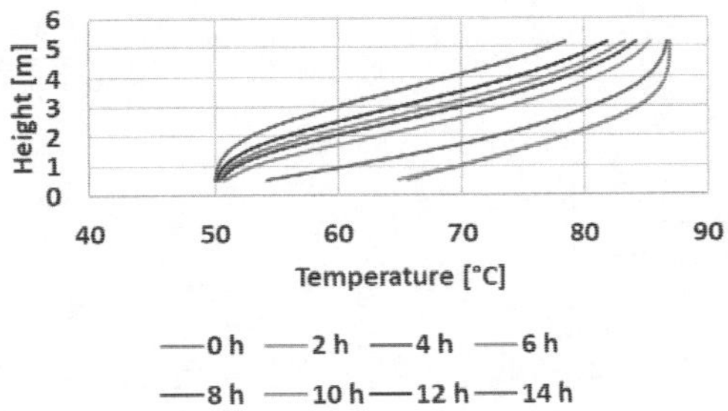

Figure 14. The temperature distribution inside the WST in the discharging (cooling) period.

The duration of the charging period is more than 5 hours, and the duration of the discharging period is more than 14 hours.

The thermal energy delivered monthly to the consumers, in the new configuration (with the CHP plant) is presented in figure 15 and the thermal energy delivered monthly by the BB is presented in figure 16.

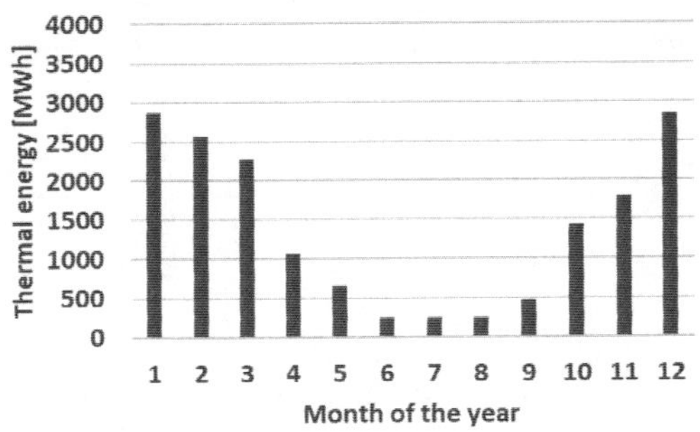

Figure 15. The thermal energy delivered monthly to the consumers.

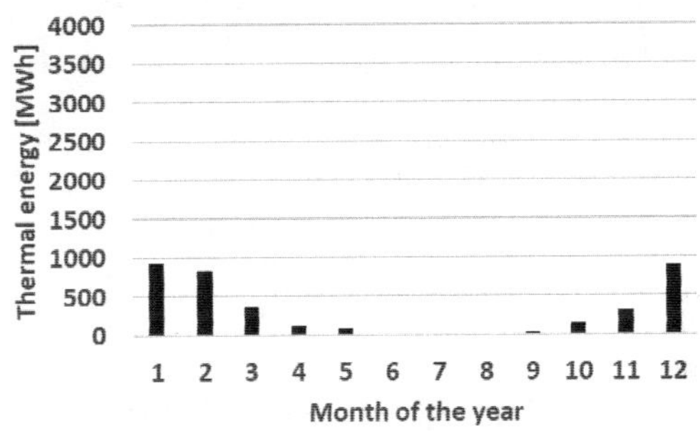

Figure 16. The thermal energy delivered monthly by the BB.

In the proposed configuration (with the CHP plant) the BB is operated, mainly to provide thermal power peaks in the winter periods.

The thermal and electric energy provided monthly by the CHP engines is presented in figures 17 and 18 respectively.

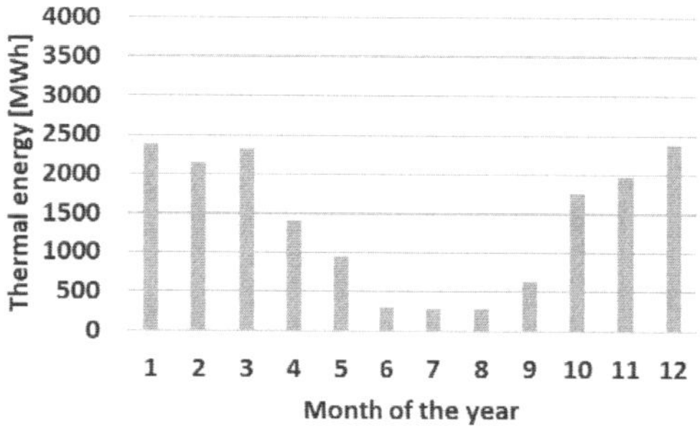

Figure 17. The thermal energy provided monthly by the CHP engines.

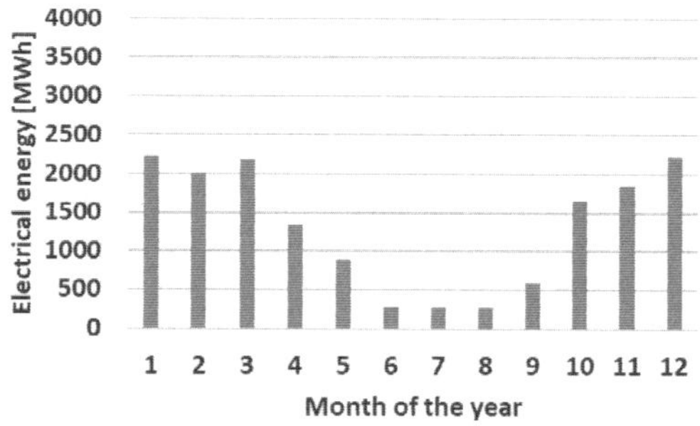

Figure 18. The electric energy provided monthly by the CHP engines.

The monthly energy delivered in and extracted from the WST is presented in figures 19 and 20 respectively.

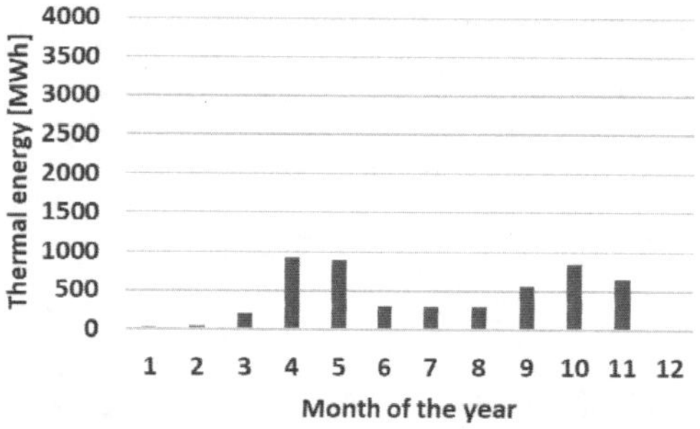

Figure 19. The thermal energy provided monthly in the WST.

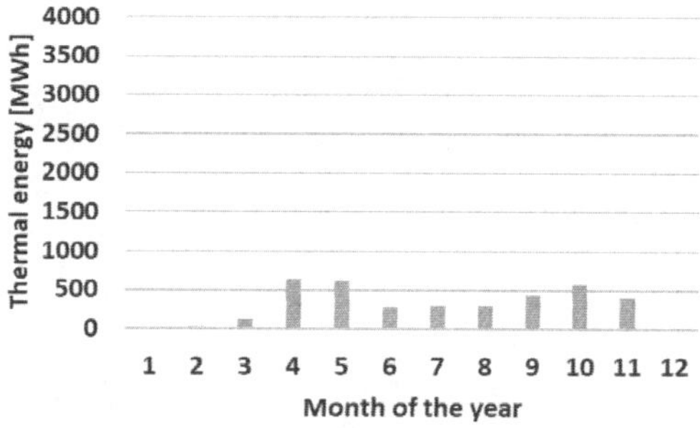

Figure 20. The thermal energy extracted monthly from the WST.

The flow rates in the distribution thermal network (from HE3 to the consumers) are provided in figure 21 and the flow rates in the transport thermal network (HE2, BB, HE3) is provided in figure 22.

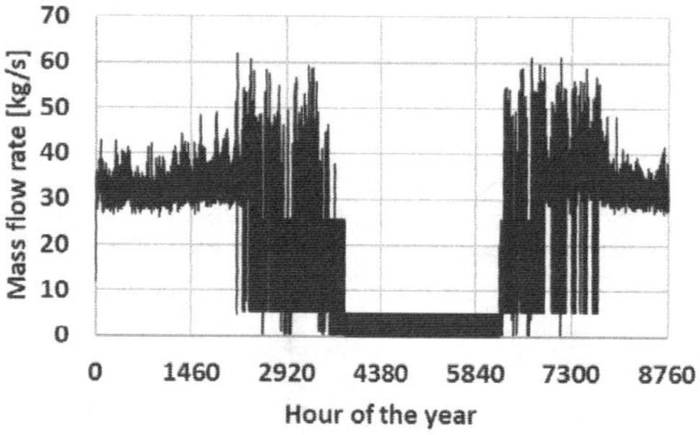

Figure 21. The flow rate in the distribution thermal network (from HE3 to the consumers).

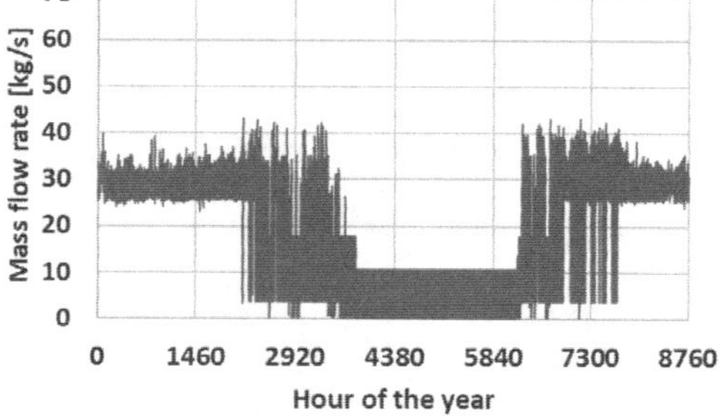

Figure 22. The flow rate in the transport thermal network (HE2, BB, HE3).

The flow rates are influenced by the transported thermal power and by the thermal regimes of the heat exchangers. Thus, in summer, when only DHW is provided, the flow rate in the distribution thermal network (from HE3 to the consumers) is lower than the flow rate in the transport thermal network (HE2, BB, HE3), because the temperature difference on HE3 is higher on the side of the distribution network, than in the transport network.

The annual heat flow Sankey diagram of the whole district heating system, in MWh, is presented in figure 23.

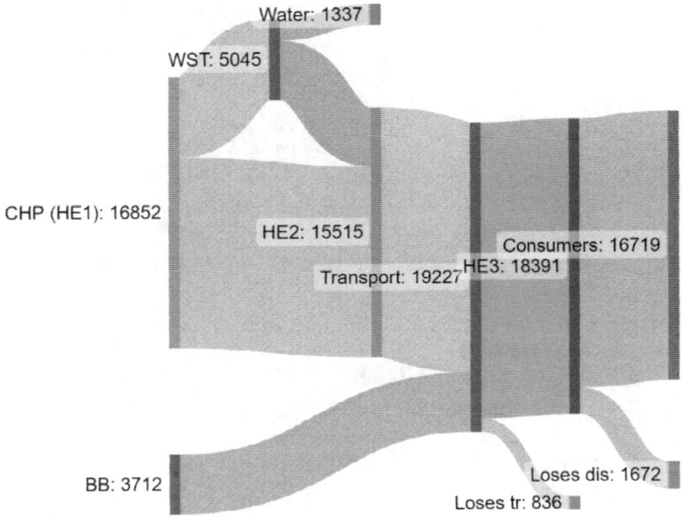

Figure 23. Sankey diagram of the heat flow in the district heating system.

The components of the annual heat flow are presented in table 4.

Table 4. The components of the annual heat flow.

From	Value [MWh]	Value [%]	To
CHP (HE1)	11807	57.4	HE2
CHP (HE1)	5045	24.5	WST
WST	1337	6.5	Water
HE2	15515	75.4	Transport network
BB	3712	18.1	Transport network
Transport network	18391	89.4	HE3
Transport network	836	4.1	Losses transport
HE3	16719	81.3	Consumers
HE3	1672	8.1	Losses distribution

The global yearly heat balance is verified:

- Produced heat: 20564 MWh
- Consumed heat (including loses): 20564 MWh

The values in percentages are computed related to the total produced heat.

IOP Conf. Series: Materials Science and Engineering 1290 (2023) 012017 doi:10.1088/1757-899X/1290/1/012017

The comparative ratios between the storage volume and thermal power of the CHP engines, for thermal powers smaller than 350 MW, are presented in figure 24.

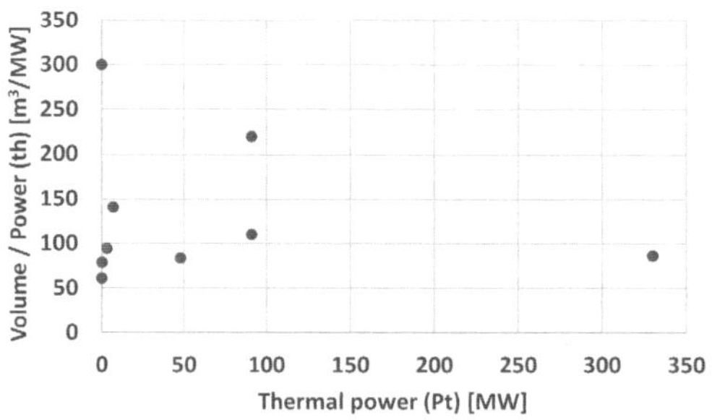

Figure 24. The comparative ratios between the storage volume and the thermal power.

The references for the parameters represented in figure 24 are provided in table 5.

Table 5. The references and the parameters provided in figure 24.

Reference	Location	Thermal power [MW]	WST volume [m^3]	Volume / Power [m^3/MW]
[3]	NA	0.2	15.7	78.5
[9]	Germany	0.0025	0.75	300.0
[9]	Germany	0.0148	0.9	60.8
[17]	Finland	91	10000	109.9
[17]	Finland	91	20000	219.8
[13]	Italy	7.1	1000	140.8
[7]	NA	48	4000	83.3
[19]	NA	330	28486	86.3
This study	Romania	3.2	300	93.8

The ratio between the storage volume and the thermal power of the CHP engines, of 93.8 corresponding to this study, and colored in red, fit in the variation range of this parameter, identified in the literature, validating that the WST was correctly sized.

4. Conclusions

The study presents the results of a numerical simulation through analytical method of the thermal behavior of a district heating located in Romania. The existing system is based on two BB and it is planned to integrate a new CHP plant based on two natural gas engines, coupled with a WST.

The WST is planned to be used mainly in summer when the heating load required only for DHW preparation is low and the thermal power of even one CHP engine is higher than this small heating load. The WST allows more flexibility in the operation of the CHP engine in the summer, allowing to synchronize the operation period with the daily intervals with high price of the electricity.

It was computed and presented the hourly and the monthly variation of many parameters, such as: temperatures and flow rates in all the characteristic states of the district heating system, temperatures distribution inside the WST, thermal and electric power provided by the CHP engines, thermal power provided by the BB, thermal power transferred through the heat exchangers, thermal power introduced in and extracted from the WST.

The total yearly heat production is 20564 MWh (81.54 % produced by the CHP plant and 18.46 % produced by the BB). The useful heat delivered to the consumers of 16719 MWh represents 81.30 % of the total yearly heat production.

The presented analysis of the district heating system, that integrates a natural gas CHP plant and a WST into an existing system based only on biomass, prove that this cohabitation is possible and presents advantages mainly due to the increased flexibility of the system. Such investigation can be successfully implemented in other district heating systems.

Acknowledgement

This study was partially financially supported by the Project "Network of excellence in applied research and innovation for doctoral and postdoctoral programs / InoHubDoc", project co-funded by the European Social Fund financing agreement no. POCU/993/6/13/153437.

This study was partially supported by a grant of the Ministry of Research, Innovation and Digitization, CCCDI - UEFISCDI, project number PN-III-P2-2.1-PED-2021-0544, within PNCDI III (PED 706).

References

[1] Fragaki A, Andersen A N and Toke D 2008 Exploration of economical sizing of gas engine and thermal store for combined heat and power plants in the UK *Energy* **33** 1659–70

[2] Streckiene G, Martinaitis V, Andersen A N and Katz J 2009 Feasibility of CHP-plants with thermal stores in the German spot market *Appl Energy* **86** 2308–16

[3] Lan T and Strunz K 2023 Droop control for district heating networks: Solution for temperature control of multi-energy system with renewable power supply *International Journal of Electrical Power & Energy Systems* **146** 108663

[4] Campos Celador A, Odriozola M and Sala J M 2011 Implications of the modelling of stratified hot water storage tanks in the simulation of CHP plants *Energy Convers Manag* **52** 3018–26

[5] Sun Y, Xu C, Xu G, Zhang H, Li B and Yang Y 2019 A comprehensive thermodynamic analysis of load-flexible CHP plants using district heating network *Int J Energy Res* **43** 4613–29

[6] Wang Z, Gu Y, Lu S and Zhao Z 2023 Optimization of thermocline heat storage tank capacity for combined heat and power plant based on environmental benefits: Scenarios for China *J Energy Storage* **57** 106303

[7] Turunen J, Majanne Y and Vilkko M 2020 Short-term Optimization of the Operation of the CHP District Heating Plant with Heat Accumulator *IFAC-PapersOnLine* **53** 13236–41

[8] Benalcazar P 2021 Optimal sizing of thermal energy storage systems for CHP plants considering specific investment costs: A case study *Energy* **234** 121323

[9] Lödige D, Thomas B and Widmann C 2015 Intelligent Management of the Heat Storage Tank for Production of Electricity on Demand Using CHP Units *Energy Procedia* **73** 239–43

[10] Bird T J and Jain N 2020 Dynamic modeling and validation of a micro-combined heat and power system with integrated thermal energy storage *Appl Energy* **271** 114955

[11] Koch K, Alt B and Gaderer M 2020 Dynamic Modeling of a Decarbonized District Heating System with CHP Plants in Electricity-Based Mode of Operation *Energies (Basel)* **13**

[12] Koch K, Höfner P and Gaderer M 2020 Techno-economic system comparison of a wood gas and a natural gas CHP plant in flexible district heating with a dynamic simulation model *Energy* **202** 117710

[13] Comodi G, Lorenzetti M, Salvi D and Arteconi A 2017 Criticalities of district heating in Southern Europe: Lesson learned from a CHP-DH in Central Italy *Appl Therm Eng* **112** 649–59

[14] Fambri G, Mazza A, Guelpa E, Verda V and Badami M 2023 Power-to-heat plants in district heating and electricity distribution systems: A techno-economic analysis *Energy Convers Manag* **276** 116543

[15] Dancker J, Götze J, Schulz F, Könneke N, Beyrau F and Wolter M 2019 Optimal design and operation of a CHP based district heating system including a heat storage and electrode boiler to increase self-consumption *2019 IEEE PES Conference on Innovative Smart Grid Technologies, ISGT Latin America 2019*

[16] Nuytten T, Claessens B, Paredis K, Van Bael J and Six D 2013 Flexibility of a combined heat and power system with thermal energy storage for district heating *Appl Energy* **104** 583–91

[17] Wang H, Yin W, Abdollahi E, Lahdelma R and Jiao W 2015 Modelling and optimization of CHP based district heating system with renewable energy production and energy storage *Appl Energy* **159** 401–21

[18] Lai F, Wang S, Liu M and Yan J 2020 Operation optimization on the large-scale CHP station composed of multiple CHP units and a thermocline heat storage tank *Energy Convers Manag* **211** 112767

[19] Wang D, Han X, Li H and Li X 2023 Modeling and control method of combined heat and power plant with integrated hot water storage tank *Appl Therm Eng* **226** 120314

[20] Haeseldonckx D, Peeters L, Helsen L and D'haeseleer W 2007 The impact of thermal storage on the operational behaviour of residential CHP facilities and the overall CO2 emissions *Renewable and Sustainable Energy Reviews* **11** 1227–43

[21] Hiris P D, Bode F, Pop O G and Balan M C 2020 Simple Modeling of the Solar Seasonal Thermal Storage Behavior *Solar Energy Conversion in Communities* ed I Visa and A Duta (Cham: Springer International Publishing) pp 21–34

[22] Hiris D P, Pop O G and Balan M C 2022 Preliminary sizing of solar district heating systems with seasonal water thermal storage *Heliyon* **8** e08932

[23] Hiris D P, Pop O G and Balan M C 2022 Analytical modeling and validation of the thermal behavior of seasonal storage tanks for solar district heating *Energy Reports* **8** 741–55

IOP Conf. Series: Materials Science and Engineering 1290 (2023) 012018 doi:10.1088/1757-899X/1290/1/012018

One degree of freedom resonance wave energy convertor

V Milkov[1], D M Dimitrov[1,*] and Sh Iliyaz[1,2]

[1] Department Mechanics and Machine Elements, Technical University of Varna, 9010 Varna, ul. Studentska 1, Bulgaria
[2] PhD student, Department Mechanics and Machine Elements, Technical University of Varna, 9010 Varna, ul. Studentska 1, Bulgaria

*corresponding author: Dm_dimitrov@tu-varna.bg

Abstract. In this work, a single degree of freedom resonant converter model is proposed and theoretical studies of the converter are carried out. The converter is suitable for waves with height of up to 2 m and period of up to 5 seconds, such as the waves on the western coast of the Black Sea. It is shown that at resonance its power increases significantly and approaches the theoretical values of the power of the swell. In high waves (up to 5 m) its efficiency decreases, but its power remains relatively constant and equal to about half of its power at resonance.

1. Introduction

Nowadays the world is shifting to renewable energy like solar, wind or sea waves energy. As a lot of technologies for solar and wind energy harvesting are commercially available, the wave energy extraction technologies are still at the stage of theoretical studies and small model testing. The review article [1] summarises the technologies and wave energy converter (WEC) designs. Authors conclude that there is no "champion" and future experimental work and optimizations will decide the construction of commercial wave energy farms.

Using the resonance conditions of the oscillating system, where large vibration amplitudes can be reached is a well-known approach for energy harvesting devices design. Many optimization studies for resonant point absorbers, suitable for various conditions of the sea waves can be found. A converter with two degrees of freedom and a linear generator are proposed in the work [2], applicable to large amplitude waves. In [3] the methodology for geometry optimization of one-body point absorber is presented. The authors apply numerical modelling in order to optimize the diameter and draft of the cylindrical buoy, using the wave data for nearshore region of Rio de Janeiro, Brazil. Point absorber with floating buoy and hydraulic power take-off system are studied in [4]. In this study a parametric optimization is carried out to find an optimal design for maximum energy harvesting in a target area off the Turkish coast of the Black Sea. The numerical analyses show that the semi-ellipsoid and other similar float profiles have superior performance compared to cylindrical float. Additional cost analysis, made by authors show that WEC with cylindrical float is the most economical due to its small draft to radius ratio.

The concept for WEC with waterproof float and build in power take-off mechanism (BI-PTO) is presented as a novel concept in [5]. The approach to build all the parts inside the outer-floater is cost effective in terms of water depth, because only slacking mooring system is needed. It is easier for repair and maintenance. The BI-PTO mechanism presented includes a spring, sliding-mass and damping

Content from this work may be used under the terms of the Creative Commons Attribution 3.0 licence. Any further distribution of this work must maintain attribution to the author(s) and the title of the work, journal citation and DOI.
Published under licence by IOP Publishing Ltd

systems, where the spring system is the most complicated. The converter is called dual resonance WEC. The scaled model of the converter is tested, and results are used for tuning a numerical model of enough accuracy.

The aim of the present study is to describe the methodology for conceptual design of single degree of freedom point absorber WEC capable to operate in resonant mode with the wave frequency on the western coast of the Black Sea, close to Bulgaria. The proposed WEC is suitable for low height waves, where the need for improving the converter efficiency is even greater.

2. Basic idea for WEC design

A cylindrical tank (buoy) immersed in water, partially filled with water, is shown on Figure 1.

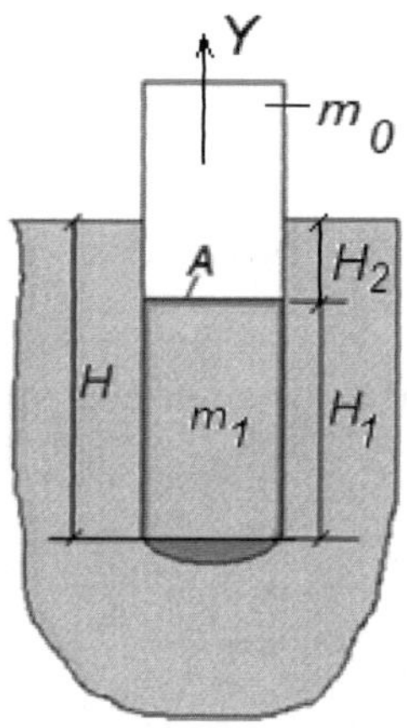

Figure 1. Cylindrical buoy

From the equilibrium condition of the submerged tank, we get:

$$(m_0 + H_1 A\rho)g = HA\rho g \tag{1}$$

where: m_0 - mass of the empty tank; ρ - density of water; g - earth acceleration; A - cross section of the tank; H - immersion depth of the tank.

Here we consider that due to the thin wall of the tank, the areas of its inner and outer cross-section are the same. From (1) it follows that:

$$H_1 = H - \frac{m_0}{\rho A} = H - H_2 \tag{2}$$

where $$H_2 = \frac{m_0}{\rho A} \tag{3}$$

Apparently H_2 is the sink depth of the empty tank.

On the other hand, the equation of the vertical free undamped oscillations of the tank partially filled with water is:

$$m\ddot{Y} + A\rho g.Y = 0 \tag{4}$$

where: Y - vertical displacement; m - total mass of the tank.

The period of vertical fluctuations will be:

$$\mathrm{T} = 2\pi/\omega = 2\pi\sqrt{\frac{m}{\rho g A}} = 2\pi\sqrt{\frac{m_0+m_1}{\rho g A}} = 2\pi\sqrt{\frac{m_0+H_1 A\rho}{\rho g A}} \tag{5}$$

where: ω - circular frequency; m_1 - mass of water in the tank; H_1- height of the water column in the tank.
In real conditions, the mass of the water joined in the oscillation must be added to the mass m_0. We solve (4) with respect to H_1. We obtain:

$$H_1 = \frac{T^2 g}{4\pi^2} - \frac{m_0}{\rho A} = \frac{T^2 g}{4\pi^2} - H_2 \quad (6)$$

Formula (6) determines the depth of water into the tank in order to create oscillations with predominant period T. We will denote the mass of water with height H_1 in the tank by m_1.
Comparing (2) and (5), one can cee that the oscillation period T is related to the immersion depth H by the expression:

$$H = \frac{T^2 g}{4\pi^2} = 0{,}249.\mathrm{T}^2 \quad (7)$$

Graphically, the dependence (7) has the form, shown on Figure.2.
If the tank is placed in the conditions of regular swell, in order to have resonance, the period of oscillation of the tank must be equal to the period of the swell.

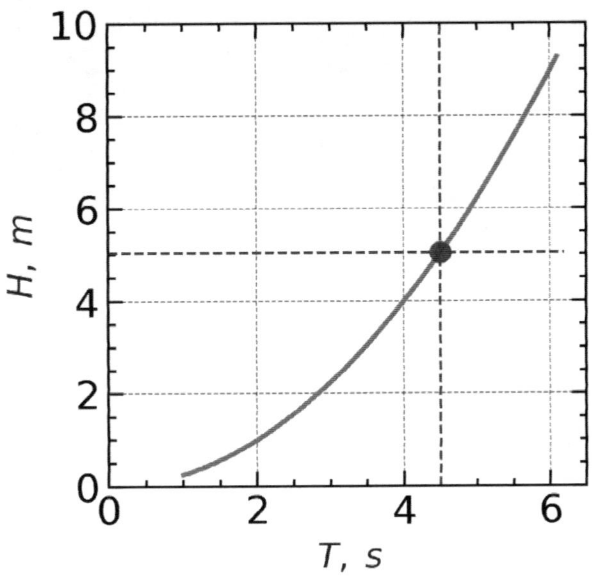

Figure 2 Dependence of buoy submergence depth H on wave period T

The graph (Figure 2) shows that for a period of 4.5 seconds (average period for waves on the western coast of the Black Sea near the coast - [5]), the height H is about 5m. Adding 2-3 m for the overwater part, we get that the tank should be about 7-8 m high. This depth does not depend on the diameter of the tank, as can be seen from formula (7). The height of the water in the tank is determined by formula (6).To improve the stability of the tank, it is appropriate to fill part of its volume with concrete and the rest with water. In this way, the centre of gravity of the tank will be lower and its metacentric height will increase. The total mass of concrete and water in the tank should equal the mass of water m_1.

3. Possible solutions for resonant converters.

Eight solutions for resonant converters are given in Figure. 3, where the direction of the working stroke is shown. For all schemes, the immersion depth is determined by expression (7). The converter in circuit 4 is a linear generator, in circuit 7 is a piston pump. The converter from scheme 5 is with a Savonius turbine and from scheme 6 is with a Wells turbine. The buoy height of all converters is 7-8 m, which allows their resonance tuning with a wave period of up to 5 s.

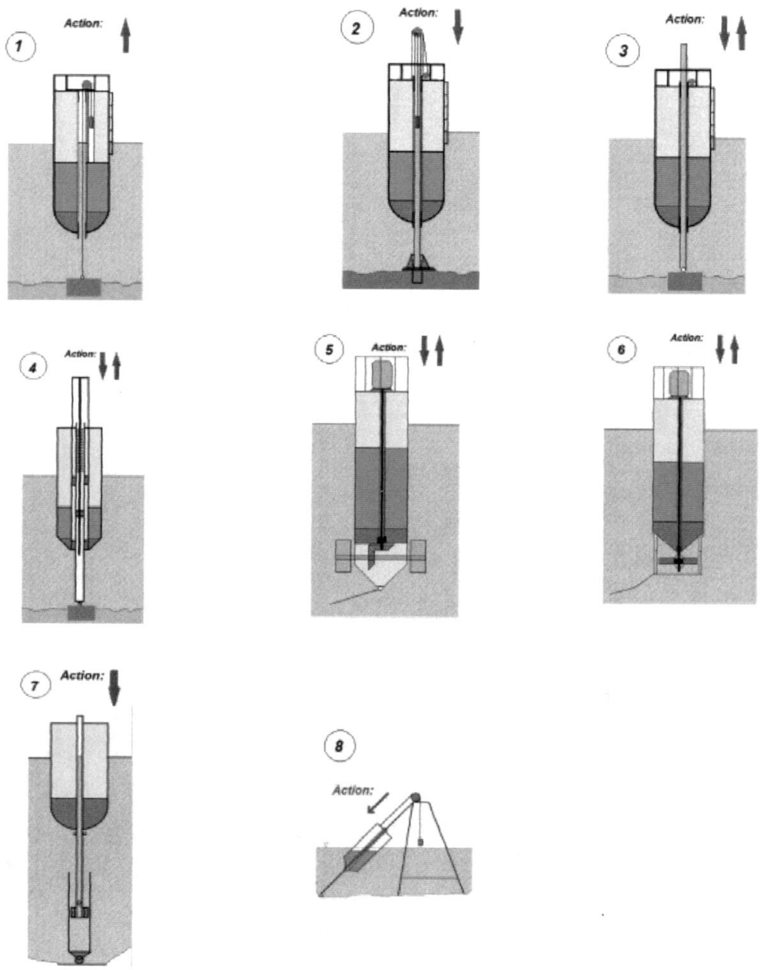

Figure 3 Various constructions of resonant WEC with rotary generator

4. Scheme of resonant WEC with rotary generator

Circuit 2 from Figure 3 will be considered in more details. The converter is a tall cylindrical tank, partially filled with seawater and concrete, Figure 4. A pylon (hollow and watertight) about 11m high is installed along the axis of the buoy, which rests on the bottom of a special heel. The buoy slides on the pylon as there are polyamide bushings at both ends of the buoy. At the upper end of the pylon a polyspast is mounted and a generator with a one-way coupling is mounted on the buoy. The polyspast is designed to increase the speed of the generator. The clutch pulley is driven by a chain, one end of which is connected to the buoy, and the other end is tensioned by a counterweight that moves inside the pylon. Because of the one-way clutch, the generator rotor turns only when the buoy is lowered down.

IOP Conf. Series: Materials Science and Engineering 1290 (2023) 012018 doi:10.1088/1757-899X/1290/1/012018

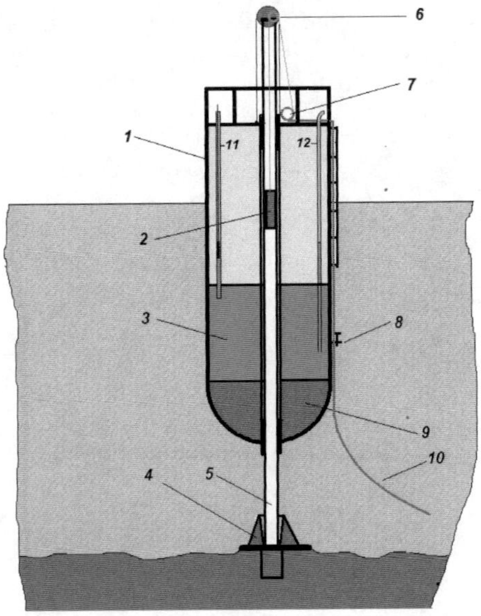

Figure 4 WEC design. 1-Buoy (cylinder), 2-Counterweight, 3-Sea water for resonance adjustment, 4-Support, 5-Pylon, 6-Polyspast, 7-Generator, 8-Faucet, 9-Concrete, 10-Cable, 11,12-Pipes

To improve the stability of the buoy, a certain amount of concrete is poured with the metacenter located above the buoy's centre of gravity. The amount of concrete and seawater in the buoy shall be such that it is tuned to the most common frequency of sea waves in the region where it is installed.

The generator has permanent magnets and is coupled to the output shaft of the multiplier. A wide range of low speed generators (min 50 rpm) are commercially available with a large choice of outputs.

The converter has the following advantages: there is no need for a heavy anchor, since during the running stroke the force generated by the generator resistance is transmitted via the pylon to the bottom; the installation of the electrical cable from the generator to the bottom is facilitated; the installation of the converter is facilitated. According to [6], converters of this type (with mechanical energy conversion) have a high efficiency, but attention must be paid to the durability of the mechanisms.

5. Differential equation of vertical oscillations

We will consider the oscillations of the converter buoy under the action of harmonic excitation, which varies according to the law:

$$Z = \frac{H_w}{2} sin\left(\frac{2\pi}{T} t\right) \tag{8}$$

where H_w - wave height; T - period of the excitement; t - time.

The period of the excitation T and the wavelength λ are related to the wave height H_w [m]. Here for the period formula (9) will be used [7, 8]:

$$T = 3.85\ H_w^{0.5}\ [s] \tag{9}$$

In principle there are six degrees of freedom for a buoy like any rigid body, but here only the vertical motion will be considered- the vertical displacement of the buoy's centre of gravity Y. The other motions affect the converter performance very little and are only loosely related to the vertical motion of the buoy. However, the vertical motion is accompanied by a swing about the buoy support in the plane perpendicular to the waves, which is defined by the Θ coordinate. This motion does not cause rotation of the generator but causes a centrifugal force along the axis of the buoy. The swing movement can be described by the expression:

$$\Theta = \Theta_a cos(\Omega t) \qquad \dot{\Theta} = -\Theta_a \Omega sin(\Omega t) \tag{10}$$

Here $\Omega = 2\pi/T$ - circular frequency of the excitation.

Taking into account that on the surface the water particles perform a rotational motion along a circle of diameter H_w, the amplitude value of the swing angle will be:

$$\Theta_a = (H_w/2)/H_d \tag{11}$$

where H_d - depth of sea water.

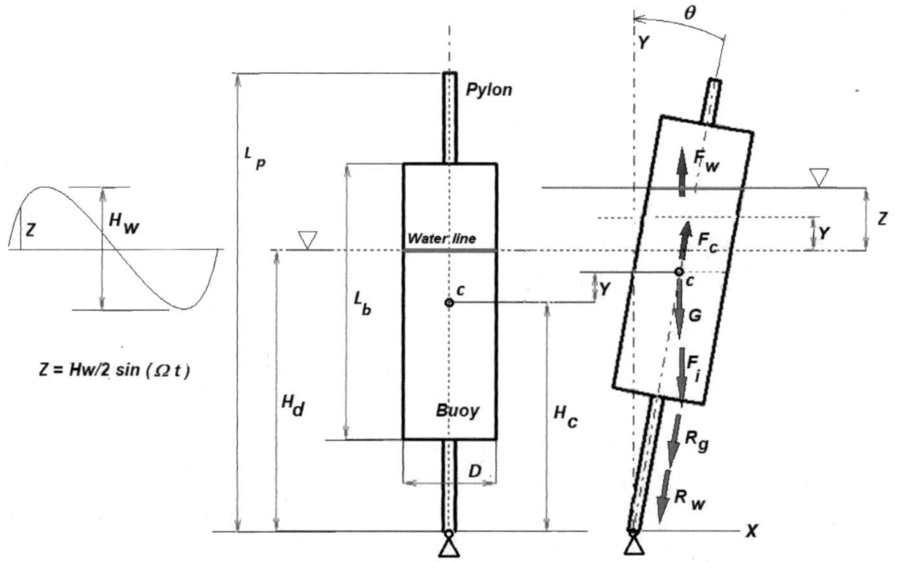

Figure 5 Forces, acting on the converter

The following forces act on the WEC buoy:

a. The force from the surge (vertical), which we get from Archimedes' law:

$$F_w = (Z - Y)A\gamma = \left(\frac{H_w}{2} sin\,(\Omega t) - Y\right) A\gamma \tag{12}$$

A – cross section area, Υ – relative weight of the water.

b. The centrifugal force from the converter swing (along the buoy axis) with angular velocity $d\Theta/dt$.

$$F_c = (m_0 + m_1)\,(Hc + Y)\,\Theta_a^2 \Omega^2 sin^2(\Omega t) \tag{13}$$

where m_0 is the mass of the converter without the pylon, water and concrete in the buoy; m_1 is the total mass of water and concrete in the buoy. This force is always directed upwards.

c. The force from the buoy's own weight (vertical):

$$G = (m_0 + m_1)\,g \tag{14}$$

d. The inertial force from the vertical movement of the buoy (vertical):

$$F_i = (m_0 + m_1 + m_a)\ddot{Y} \qquad (15)$$

Here, m_a is the mass of added water in the vertical oscillations of the buoy, which depends on the shape of the buoy.

f. The resistance force from the movement of the buoy in the water (along the axis of the buoy):

$$R_w = \frac{\rho \dot{y}^2}{2} A.C_z \qquad (16)$$

Here we use Bernoulli's law by assuming that the drag is a square function of the buoyancy velocity. The drag coefficient C_z depends on the shape of the bottom. For hemisphere C_z= 0.34.

g. The resistance force from the operation of the generator (along the axis of the buoy):

$$R_g = M_g / R_{cluch} \qquad (17)$$

Here M_g - resistance moment at the input of the generator rotor (sea 21) ; R_{cluch} – radius of the one-way clutch pulley.

Because of the small swing angle, we will assume that all forces are vertical. From the condition for equilibrium of forces we obtain (neglecting the force from self-weight, which is constant and does not affect the oscillations):

$$F_i + R_g + R_w = F_w + F_c \qquad (18)$$

As a result, the following differential equation is obtained:

$$\ddot{Y} = \left(F_w + F_c - R_g - R_w\right)/M_{con} \qquad (19)$$

where: M_{con} – total mass of the converter.

To this equation the differential equation of the rotation of the generator rotor is added:

$$\dot{\omega} = -\mathrm{M}_g/J_g \qquad (20)$$

where: ω – angular velocity of the generator rotor; Jg – mass moment of inertia of the generator rotor.

For the resistance moment of the generator, the following expression is found:

$$\mathrm{M}_g = 91{,}2k^2\,\omega/R \qquad [\mathrm{Nm}] \qquad (21)$$

Where: R – generator load resistance (Ohm);

$$k = \frac{U_{nom}}{n_{nom}} \qquad (22)$$

Here U_{nom} , n_{nom} are nominal voltage (V) and nominal revolutions (min^{-1}) of the generator.

Here we will note that when the buoy moves down, the angular velocity of the rotor is equal to the angular velocity of the clutch, and when the buoy moves up, the rotor is decoupled by the one-way clutch and the rotor rotates by inertia.

The system of differential equations (19) and (20) is non-linear and can only be solved numerically. It is reduced to a system of three differential equations of the first order with initial conditions and is solved by known methods, for example Runge-Kutta of the fourth order.

6. Numerical example

For the numerical solution of the differential equations a software is developed. Numerous analyses of various WECs have been performed with the software program developed form the first author. For a converter with a 10kW generator (buoy diameter 2m and height 6m, detailed data are given in Figure.

IOP Conf. Series: Materials Science and Engineering 1290 (2023) 012018 doi:10.1088/1757-899X/1290/1/012018

6.) A sample of some results are shown on Figure 7. The converter is set for a wave with a height of 1.3m and a period of the wave *T= 4.4 s*, which wave parameters are common on the western coast of the Black Sea -[9, 10]. For this purpose, 6 tons of concrete and 5 tons of seawater were poured into the buoy - a total of 11 tons of additional mass. The average power is about 7 kW at 96% efficiency. The efficiency is calculated as the ratio of the received power and the theoretical power of the waves at the set parameters according to the linear wave theory [8].

$$P_{Theor} \approx 0,5\, H_w^2.\, T.\, D \tag{23}$$

If the buoy is filled with only 2-3 tons of concrete, the converter will operate with a power of about 2-3 kW.

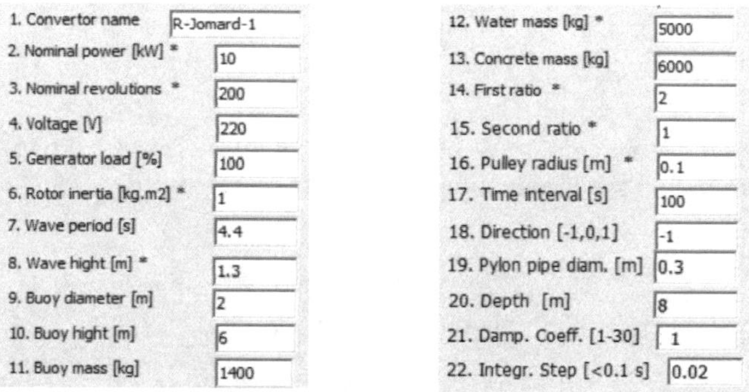

Figure 6. Proposed WEC input data

Depending on the season, the parameters of the excitement change within very wide limits. The questions how the converter will behave at a wave height of up to 5 m, what waves are possible in the Black Sea [11] arise. Figure 7 and Figure 8 show the power variation, maximum displacement and converter efficiency as a function of the wave height.

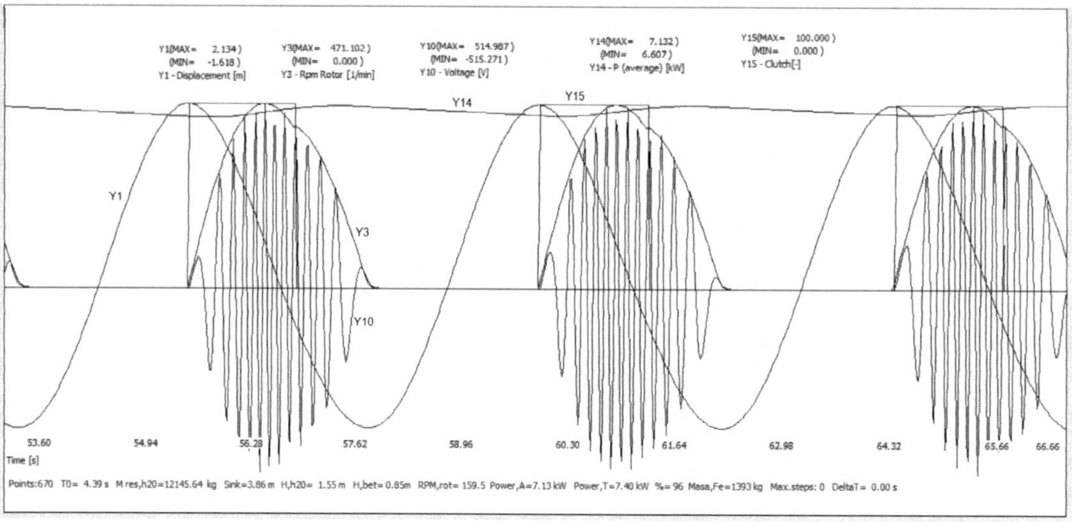

Figure 7. Sample results for a 10 kW converter: Y_1 - buoy displacement, Y_3 - generator speed, Y_{10} - generator voltage, Y_{14} - average generator power, Y_{15} - one-way clutch status (on/off).

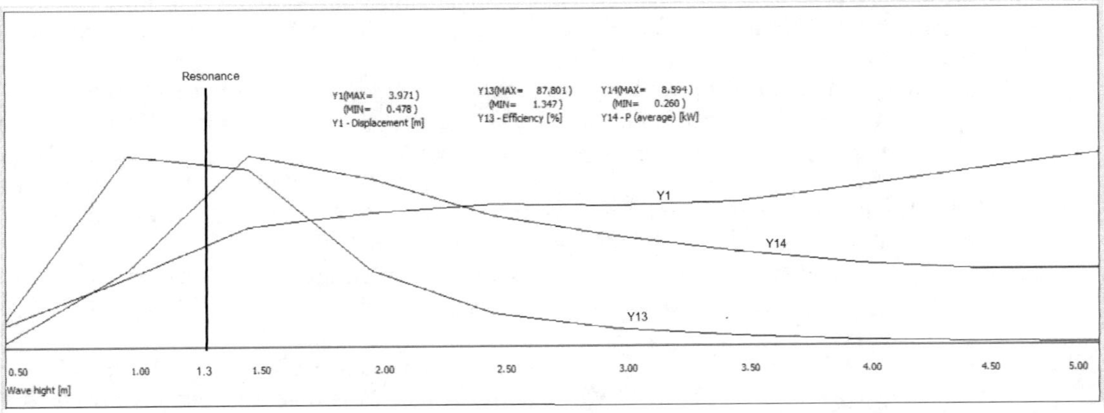

Figure 8. Dependence of the amplitude Y_1, power Y_{14} and efficiency Y_{13} of the converter on the wave height

It can be seen from figures that the maximum efficiency of the converter is at a wave height of 1.3m (resonance), where its power is significantly higher than the off-resonance power. A very important feature is that at a high wave height (up to 5 m) its efficiency greatly decreases, but its power is relatively constant and equal to about half of the maximum power, and the displacement of the buoy increases slightly, compared to the displacement at resonance.

7. Conclusions

The proposed WEC and the conducted numerical study allow the following conclusions to be drawn:

- The converter is suitable for waves with a height of up to 1.5 m and a wave period of about 4-5 s, such waves are often found on the west coast of the Black Sea. With these parameters of the waves, the converter has a height of 6-7 m. Its power depends linearly on the diameter.

-The converter can be tuned to operate in a resonant mode with the frequency of the waves. At the same excitation parameters, the efficiency of the converter at resonance increases by more than 50%.

-The converter is not suitable to be set to resonance for large waves (above 2 m), because due to increase in the period of the waves, the converter becomes very high. For example, with a wave height of 2 m, the converter will be about 10 m high.

-The power of the converter after resonance drops by half, but remains almost constant even at high waves (up to 5 m). All other parameters of the converter - displacement, speed, revolutions of the generator, etc. are increased slightly from their values at resonance, which prevents it from damage.

References

[1] Aderinto T and Li H. 2018. Ocean Wave Energy Converters: Status and Challenges. *Energies* **11**(5):1250. https://doi.org/10.3390/en11051250
[2] Engstro¨m V. Kurupath J. Isberg, and M. Leijon. 2011. A resonant two body system for a point absorbing wave energy converter with direct-driven linear generator. *Journal of applied physics* **110**, 124904
[3] Shadman, Milad, et al. 2018. A geometrical optimization method applied to a heaving point absorber wave energy converter. *Renewable energy* 115: 533-546.

[4] Erselcan İ Ö, and Abdi Kükner. 2020. A parametric optimization study towards the preliminary design of point absorber type wave energy converters suitable for the Turkish coasts of the Black Sea. *Ocean Engineering* **218**: 108275.

[5] Chen, Zhongfei, et al. 2018. Experimental and numerical study on a novel dual-resonance wave energy converter with a built-in power take-off system. *Energy* **165**: 1008-1020.

[6] Arthur Pecher, Jens Peter Kofoed. *Handbook of Ocean Wave Energy.* Springer Open. 2017

[7] Kyung-Duck Suh, Hyuk-Dong Kwon, Dong Young Lee. 2010. Relationship Between Significant Wave Height and Period and Some Spectral Parameters of Large Deepwater Waves. *Proceedings of the 17th IAHR APD Congress* (Auckland, 2010).

[8] Goda Y. 2000. Random Seas and Design of Maritime Structures. Advanced Series on Ocean Engineering. **15**. *Singapore: World Scientific Publishing Company*. ISBN 978-981-02-3256-6.

[9] Grozdev D. 2006. Wind wave climate along the Bulgarien Black Sea coast. *GEOSCIENCES*

[10] Maria Emanuela Mihailov , Luminita Buga, Sabina Stefan, Maria Ionela Tomescu-Chivu, Petrica Popov, and Lucian Dumitrache. 2013. Waves and Marine Currents Characteristics Along the Western Black Sea Coast . *Geophysical Research Abstracts* **15**, EGU2013-9161. EGU General Assembly 2013.

[11] Loránd Sz and Claudiu O. 2006. Linear Generators for Wave Power Plants to Be Set up Near the Romanian Coasts of the Black Sea. *Department of Electrical Machines, Technical University of Cluj*. Oradea.

AUTHOR INDEX